T0256350

Digital Signals Theory

Where most introductory texts to digital signal processing assume a degree of technical knowledge, this class-tested textbook provides a comprehensive introduction to the fundamentals of digital signal processing in a way that is accessible to all.

Beginning from the first principles, readers will learn how signals are acquired, represented, analyzed, and transformed by digital computers. Specific attention is given to digital sampling, discrete Fourier analysis, and linear filtering in the time and frequency domains. All concepts are introduced practically and theoretically, combining intuitive illustrations, mathematical derivations, and software implementations in Python. Practical exercises are included at the end of each chapter to test readers' knowledge.

Written in a clear and accessible style, *Digital Signals Theory* is particularly aimed at students and general readers interested in audio and digital signal processing but who may not have extensive mathematical or engineering training.

Brian McFee is an assistant professor of music technology and data science at New York University. His research lies at the intersection of machine learning and audio analysis, with a specific focus on music information retrieval. In addition, he is an active open-source software developer and the principal maintainer of the Librosa package for audio analysis.

Digital Signals Theory

Brian McFee

CRC Press

Taylor & Francis Group

Boca Raton London New York

CRC Press is an imprint of the
Taylor & Francis Group, an **informa** business

A CHAPMAN & HALL BOOK

First edition published 2024
by CRC Press
6000 Broken Sound Parkway NW, Suite 300, Boca Raton, FL 33487-2742

and by CRC Press
4 Park Square, Milton Park, Abingdon, Oxon, OX14 4RN

CRC Press is an imprint of Taylor & Francis Group, LLC

Library of Congress Cataloging-in-Publication Data

Names: McFee, Brian, author.
Title: Digital signals theory / Brian McFee.
Description: First edition. | Boca Raton, FL : CRC Press, [2024] | Includes
bibliographical references and index. | Summary: "This class-tested
textbook is intended for students interested in learning digital signal
processing from the ground up, but who may not have much mathematical or
engineering training. The focus is on digital signals, meaning discrete
time signals as represented in modern computers. Unlike many other
books, this book does not cover continuous time signals (except insofar
as necessary to understand digital sampling). The scope of the book is
limited to these general topics: Signals and systems, Sampling theory,
Discrete Fourier analysis, and Discrete-time linear filtering"--
Provided by publisher.
Identifiers: LCCN 2023013367 (print) | LCCN 2023013368 (ebook) | ISBN
9781032207148 (hbk) | ISBN 9781032200507 (pbk) | ISBN 9781003264859
(ebk)
Subjects: LCSH: Signal processing--Digital techniques.
Classification: LCC TK5102.9 .M395 2024 (print) | LCC TK5102.9 (ebook) |
DDC 621.382/2--dc23/eng/20230811
LC record available at https://lccn.loc.gov/2023013367
LC ebook record available at https://lccn.loc.gov/2023013368

ISBN: 978-1-032-20714-8 (hbk)
ISBN: 978-1-032-20050-7 (pbk)
ISBN: 978-1-003-26485-9 (ebk)

DOI: 10.1201/9781003264859

Typeset in Latin Modern font
by KnowledgeWorks Global Ltd.

Publisher's note: This book has been prepared from camera-ready copy provided by the authors.

For Liane and Ian

Contents

Preface

This book is designed to follow the course syllabus of *Fundamentals of Digital Signals Theory (I)* (MPATE-GE 2599) at New York University. The focus here is on *digital* signals, meaning discrete-time signals as represented in modern computers. Unlike many other books, we do not cover continuous-time signals, except insofar as necessary to understand digital sampling. The scope of the book is limited to these general topics:

- Signals and systems,

- Sampling theory,

- Discrete Fourier analysis, and

- Discrete-time linear filtering.

While certainly not a comprehensive treatment of signal processing, the topics covered here should provide a solid foundation upon which readers can develop in whichever direction they see fit.

Who is this book for?

This book is intended for students interested in learning digital signal processing from the ground up, but who may not have much mathematical or engineering training. Because we do not cover the continuous-time case, we will not need differential calculus. In some places we'll have to gloss over a couple of technical details, but I hope students can still gain a sufficiently rich understanding of digital signals with minimal mathematical background.

I've tried to make the contents of this book self-contained, and provide supplementary background material in the Appendix. That said, we do have to start somewhere, and I generally expect readers to have familiarity with high-school-level algebra and geometry.

Why another book?

Put simply, I wasn't happy with any of the existing digital signals textbooks that could be used for this class. While many of them are excellent reference books, they often assume a fairly sophisticated technical background, and are aimed at upper-division undergraduate students in engineering programs. After stubbornly trying to make do with existing books, I got frustrated enough to make my own!

Supplemental resources

This book is meant to be fairly self-contained, but it certainly can't cover everything. Depending on your background, you may find yourself wanting more foundational material (e.g., on mathematics or programming), or more advanced coverage of certain topics.

Mathematics

- Khan Academy - Algebra is a good resource if you want to brush up on basic algebra.

- 3Blue1Brown has several excellent videos on topics related to this material:

    - Understanding $e^{i \cdot \pi}$ in 3.14 minutes
    - A visual introduction to the Fourier transform

Programming

- PythonTutor is an excellent resource to learn the basics of program operation.

- NumPy for MATLAB® users is helpful if you have prior experience with MATLAB, and want some basic recipes to translate into Python.

- The Software Carpentry organization has excellent tutorials for programming in Python:

    - Programming with Python
    - Plotting and Programming in Python

Further reading

- *Mathematics of the Discrete Fourier Transform* by Julius O. Smith III.

- *Understanding Digital Signal Processing* (3rd ed.) by Richard Lyons [Lyo04].

- *Discrete-Time Signal Processing* by Alan Oppenheim and Ronald Schafer [Opp10].

- *Fundamentals of Music Processing* by Meinard Müller [Mul15].

Acknowledgments

I would like to thank several people for providing feedback on early versions of this text: Meinard Müller, Frank Zalkow, Ernesto Valenzuela, Haokun Tian, Katherine Kinnaird, Tanya Clement, and Nancy Rico Mineros.

This project would not have been possible without the tireless efforts of the open source software developers, especially contributors to the following projects:

- NumPy

- SciPy

- Matplotlib

- Jupyter

- Jupyter-Book

- Sphinx

- Freesound

- Soundfile

- libsndfile

## CHAPTER 1

Signals

This chapter introduces the concept of signals as mathematical objects. We'll first establish some preliminaries and notation, then give an overview of waves and their role in signal analysis. We'll finish with a quick discussion of units and dimensional analysis and a high-level introduction to acoustics.

1.1 PRELIMINARIES

Before we get too far into digital signals, we'll need to establish some basic notation and concepts. This section covers the following:

- What is a signal?

- Mathematical notation

- Standard conventions

It does not cover mathematical fundamentals. For the first few chapters, we will not need much math beyond basic algebra (variables and rules of arithmetic) and a bit of geometry. Later on, we'll make use of more advanced concepts (complex numbers, polynomials, and exponentials), but those will be introduced as needed.

Note. I sometimes use the term *basic* when describing certain ideas. By this, I mean that these ideas are the *base level* upon which we will build. This is **not** the same thing as being *easy*. Many basic ideas can be complicated and take quite some time to thoroughly understand.

If the mathematical concepts start to seem overwhelming at some point, **don't worry**. Just take it slow and don't move on to the next section too soon. None of the contents here are magical: it all builds on basic ideas.

DOI: 10.1201/9781003264859-1

1.1.1 What is a signal?

At a high level, a **signal** is a way of conveying information. There unfortunately isn't a simple, concise, technical definition of *signal* that provides us with much insight. The most intuitive example of a signal, at least for acoustically inclined people, is the voltage on a wire connected to a microphone. If you were to watch the voltage change over time, and you might see something like Figure 1.1.

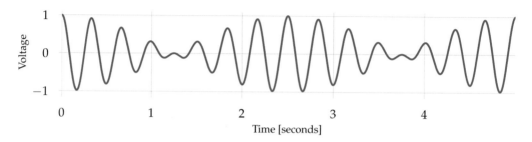

Figure 1.1 An example signal showing voltage changing over time.

The blue curve above represents the voltage measured at every time t, and we denote this curve (our signal) by the notation

$$x(t).$$

If we want to know the voltage at a specific time, say, 4 seconds, we would substitute that value for t to get $x(4)$. (In this case, $x(4) \approx 0.3$.) Not all signals measure voltages, though at least recorded audio signals are typically constructed from electrical signals. For audio processing, the important thing is that the signal can be converted somehow into changes in air-pressure that our ears can detect and interpret as sound. Whether $x(t)$ is measuring volts, Pascals (the standard unit of air pressure), or something else, is usually not too important, as we are primarily concerned with the relative changes in the signal and not its absolute interpretation. For that reason, we usually do not bother precisely annotating the units of whatever the signal is measuring.

To be a bit more precise, the signal $x(t)$ above is a *time-domain* signal. There are other kinds of signals: for example, a gray-scale image can be thought of as a signal over two input coordinates (horizontal and vertical position), and the measurement is the brightness at that point. Most of the techniques we develop in this book can be carried over to more general types of signals, but we will focus on the time-domain.

1.1.2 Notation

The first bit of notation that we'll need is for signals.

- $x(t)$ denotes a **continuous-time** signal. t can be any real number: 0, 1, -53, π, etc. We read this as "signal x at time t."

- $x[n]$ denotes a **discrete-time** signal. n must be an integer: $0, 1, 2, -1, -2$, etc. We read this as "the nth sample of signal x."

For this first chapter, we'll deal only with continuous signals $x(t)$. Discrete signals will be introduced in the next chapter, but it's important to keep clear the distinction of notation: parentheses for continuous time, square brackets for discrete sample indices. When the usage is clear from context, we may omit the time or sample index and refer to a signal just by its name x – this should be understood as referring to the entire signal.

Because we often use the letter "x" to refer to signals, we'll need a different symbol to represent multiplication. We'll instead use the center dot · for this, e.g., $2 \cdot 3 = 6$. This creates a bit of a disconnect between multiplication in software (usually denoted by *) and in text, but this is for a good reason: the * operator will be used for a different purpose later on, known as *convolution*. Using · instead of * for multiplication will help avoid confusion due to notation issues.

Usually, we'll let $x(t)$ represent an arbitrary signal. If we need to refer to a specific signal, it will be defined explicitly by a mathematical formula, such as this example

$$x(t) = \sin(2\pi \cdot t)$$

would define a pure tone oscillating at 1 cycle per second.

Most of what we do in signal processing amounts to modifying a signal in some way, for example, applying a low-pass filter to remove high-frequency content. We can think of this as applying some **function** g to an **input signal** x and producing an **output signal** y, which we write as

$$y = g(x).$$

1.1.3 Standard conventions

We'll often use $x(t)$ and $y(t)$ to generally refer to input and output signals, respectively.

To keep things simple, time will always be referred to by the letter t, and it will always be measured in units of seconds. As stated above, time should be understood to be continuously valued. Time will be measured starting from 0, as illustrated in Figure 1.2.

Discrete (i.e., integer-valued) quantities will be referred to by the letters n, m, or k. These also will be measured starting from 0. This means that a sequence of length N has indices $0, 1, 2, \ldots, N-1$. If you're not used to this, it can seem a bit awkward at first, but trust me: this convention will make the math much easier later on.

Unless otherwise stated, we will by default assume that signals are 0-valued when t (for continuous-time) or n (for discrete-time) are negative:

$$
\begin{aligned}
t < 0 &\Rightarrow x(t) = 0 \\
n < 0 &\Rightarrow x[n] = 0.
\end{aligned}
\tag{1.1}
$$

You can think of this as pretending that signals are silent until you start recording.

Angles will be denoted by the Greek letters θ (*theta*) or ϕ (*phi*), and measured in units of *radians*, **not degrees**, as illustrated in Figure 1.3. There are 2π radians in a full rotation ($360°$); π radians in a half rotation ($180°$), and $\pi/2$ radians in a right angle ($90°$).

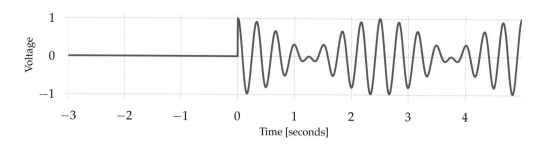

Figure 1.2 An example signal showing that voltage $x(t) = 0$ for all $t < 0$ (negative time).

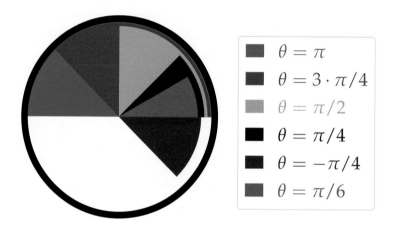

Figure 1.3 Illustrations of angles measured in radians, denoted by θ.

We'll use the letter j to refer to the *imaginary unit*, i.e., $j = \sqrt{-1}$. Mathematicians traditionally use i for this, but j is more commonly used in electrical engineering because i is reserved for electrical current. Fortunately for us, the Python programming language also uses `1j` to denote the imaginary unit, so translating from text to code should be fairly straightforward.

Complex numbers – numbers with both a real and imaginary component – will be denoted generally by the letter z (or occasionally w). This is primarily for consistency with other mathematical traditions. In general, x and y can assumed to be real-valued, and z can be assumed to be complex-valued unless otherwise stated.

Finally, we will occasionally see snippets of code that implement the abstract concepts being described in the text. This code will be written in the Python programming language and will be typeset as follows:

```python
for n in range(len(x)):
    y[n] = 2*x[n]   # Hurray! Now y is twice as big as x
```

More specifically, I will assume Python 3.6 or later and will make use of the `NumPy` package for numerical computation. If you're new to Python, I recommend getting started with the Anaconda distribution.

1.2 PERIODICITY AND WAVES

This section introduces the concept of *periodicity*, which characterizes repeating signals. This idea is central to many concepts in both signal processing and perception of audio, so it's worth spending some time with.

Specifically, this section covers the following topics:

- Periodicity

- Fundamental frequency

- Waves (sinusoids) and their parameters

- Basic properties of waves

1.2.1 Periodicity

Definition 1.1 (Periodicity). A signal $x(t)$ is said to be *periodic* if there exists some finite $t_0 > 0$ such that for every time t,

$$x(t + t_0) = x(t).$$

The smallest such t_0 satisfying this equation, if it exists, it is called the *fundamental period* (or sometimes just *period*) of the signal x. Different signals may have different periods, and some signals may have no period at all.

Think of the period of a signal as the shortest amount of time it takes, when listening to a looped recording, before you hear the recording start over.

Example 1.1 (Pulse train). A *pulse* train is a special kind of signal that consists only of ones and zeros, with the ones being separated at regular intervals. An example pulse train with a spacing of 1.5 seconds between pulses can be defined mathematically as

$$x(t) = \begin{cases} 1 & \text{if } t \in 0, 1.5, 3, 4.5, \ldots \\ 0 & \text{otherwise} \end{cases} \tag{1.2}$$

and the first few seconds are visualized in Fig. 1.4.

Fig. 1.4 demonstrates the pulse train which has a repeating pulse every 1.5 seconds. The signal is periodic with $t_0 = 1.5$ because every point on the blue curve is identical to the point exactly t_0 seconds later. Note that this must hold for **every** time t, not just the locations of the pulses! The arrows indicate two such repetitions, though in general there will be infinitely many.

This points to a key consequence of periodicity.

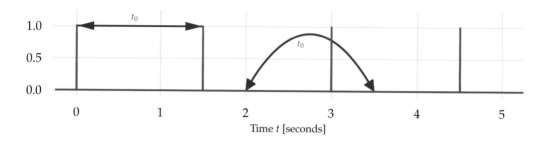

Figure 1.4 A pulse train $x_\perp(t)$ with a period of $t_0 = 1.5$ seconds.

Tip. If a signal is periodic, it must repeat forever:

$$x(t) = x(t + t_0) = x(t + 2 \cdot t_0) = x(t + 3 \cdot t_0) = \ldots$$

This is why we take t_0 to be the smallest possible period.

1.2.2 Aperiodic signals

If a signal does not have a period, we call it *aperiodic*, and assign period $t_0 = \infty$. Think of this as meaning that you would have to wait (literally) forever before you see the signal repeat itself exactly.

Aperiodic signals come in all kinds of flavors, as depicted in Figure 1.5. Random noise (e.g., white noise) is aperiodic, and if it wasn't, we probably wouldn't call it noise. However, don't think that all aperiodic signals are random or unpredictable. For example, a monotonically increasing sequence $x(t) = t$ is perfectly predictable, but it is also aperiodic because you never see the same value twice.

As a final example, a staircase signal, which is like the monotonically increasing sequence but is piece-wise constant, is also aperiodic even though you do see the same values multiple times.

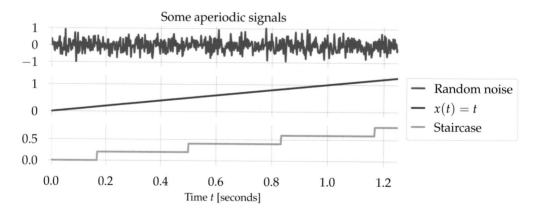

Figure 1.5 Examples of aperiodic signals. In each case, there is no finite time $t_0 > 0$ where $x(t + t_0) = x(t)$.

There's a real sense in which most signals are aperiodic. We typically use periodicity as a conceptual tool to understand idealized signals, but it's also helpful to remember that any signal of finite duration can be made periodic by playing it on a loop. This idea will come back later when we get to the Fourier transform.

1.2.3 Fundamental frequency

Periods are defined in units of time: how long must we wait before observing the signal repeating itself? Often, it is more convenient to think in terms of *how many times* a signal repeats itself within a fixed duration of time, typically one second.

This idea provides a definition for **frequency**: how many cycles does a signal complete in one second? Frequency is measured in units of Hertz (**Hz**), where 1 Hz denotes one full cycle in one second.

Definition 1.2 (Fundamental frequency). If a signal $x(t)$ has a fundamental period t_0, then its **fundamental frequency** is defined as

$$f_0 = \frac{1}{t_0}.$$

Note that f_0 need not be a whole number. In the *pulse train example* above, the period was $1.5 = 3/2$ seconds, which is equivalent to $f_0 = 2/3$ Hz: it completes two cycles in three seconds. (The third cycle starts at $t = 3$, as illustrated in the figure.)

If a signal is aperiodic and has $t_0 = \infty$, then its fundamental frequency is defined to be $f_0 = 0$, meaning that it completes 0 cycles in 1 second or any number of seconds. Note that this definition is primarily used for notational consistency – if a signal is aperiodic, it is also common to say that it has no fundamental frequency.

Why the t_0 and f_0 business?

At this point, you might (rightly) be wondering why we have these 0 subscripts on "fundamental" values. There are two reasons for this:

1. To distinguish fundamental period (t_0) or frequency (f_0) from arbitrary times (t) or frequencies (f), and

2. To highlight the connection between the fundamental f_0 and "harmonics" or "overtones", which are often notated as $f_1, f_2, \cdots$.

The first point is a somewhat arbitrary convention, but the second point is definitely not arbitrary. However, it's important to keep in mind that any periodic signal will have a fundamental frequency, but we typically only discuss harmonics in the context of sinusoids.

This naturally leads us to the question: what is a wave?

1.2.4 Waves

So far, we've discussed arbitrary signals in terms of their periodicity properties and seen examples of periodic and aperiodic signals. There are many kinds of periodic signals: pulse trains, square waves, triangle waves, sawtooth waves, just to name a few. Among all periodic signals, **sinusoids** (e.g., sine and cosine waves like Figure 1.6) are, in many ways, the most important and mathematically well-behaved. The term *wave* is, therefore, often used synonymously with the more specific *sinusoid*.

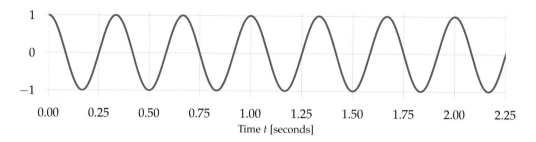

Figure 1.6 A sinusoid at 3 Hz.

What makes sinusoids special?

While many readers with some prior familiarity with audio have an intuitive grasp of sinusoids as "pure tones", it's often not clearly stated **why** sinusoids are so well-suited to signal processing and analysis, or where they come from in the first place. The reasons underlying the use of sinusoidal waves go mathematically deep, but the basic principles can be understood directly in terms of geometry.

Keep in mind that our goal here is to understand periodicities in signals. The physical universe readily supplies us with countless examples of periodic phenomena: think about the rotation of the Earth (day and night cycles), the orbit of the Moon around the Earth (full and new Moon cycles), the Earth orbiting the sun, and so on. Each of these phenomena are *repetitive*. Moreover, each of these phenomena are characterized by continuous **rotation**: once a full rotation has been completed, the cycle repeats itself.

Put succinctly, **rotation models repetition**.

Rotation and sinusoids

Sinusoids are typically introduced in a high-school geometry class in the study of right triangles. The sine of an angle θ is defined as the ratio of the opposite side-length to the hypotenuse; the cosine being the ratio of the adjacent side-length to the hypotenuse, and so on. While it is certainly useful in many contexts, this view can obscure the interpretation of sinusoids as *waves*.

Here, we'll use a different, but equivalent definition of the sine (and cosine) of an angle. First, we will draw a circle with radius 1 centered at the origin ($x = y = 0$). Next, we will draw a line through the origin and making an angle θ with the horizontal axis. This line will intersect the circle at a point, as illustrated in Figure 1.7. Then:

- $\sin(\theta)$ is the height (distance up from the horizontal axis) of the point on the circle at angle θ;

• $\cos(\theta)$ is the width (distance right from the vertical axis) of the point on the circle at angle θ.

If the circle has a radius $R \neq 1$, then these quantities would scale accordingly: $R \cdot \sin(\theta)$ and $R \cdot \cos(\theta)$ for the vertical and horizontal positions of the point on the circle, respectively.

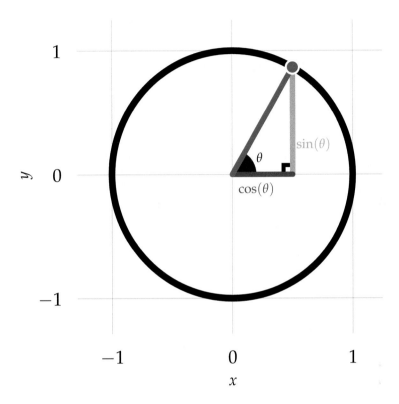

Figure 1.7 The sin and cos of an angle give the height and width of a right triangle whose hypotenuse connects the origin $(0, 0)$ to a point on the unit circle.

Note that when the circle has a radius of 1, the hypotenuse will always have length 1, so sin and cos measure the height and width of the triangle.

By convention, we take $\theta = 0$ to be the right-most point on the circle, the point $(x = 1, y = 0)$. Positive angles $\theta > 0$ correspond to a **counter-clockwise** (upward) rotation from $(1, 0)$. Negative angles $\theta < 0$ correspond to a **clockwise** (downward) rotation.

Remember: **sines and cosines turn angles into distances**.

Why does this give us waves?

The sine or cosine of a single angle just gives us a single number, between -1 and +1. A single number is not enough to get a wave – for that, we'll need to change the angle over time. Imagine the angle varying continuously over time, like the seconds hand of a clock. We'll denote this by $\theta(t)$, and have $\theta(0) = 0$ indicate that the starting position is at the right-most point on the circle $(1, 0)$.

Equivalently, we can think of this changing $\theta(t)$ in terms of the (x, y)-position of the point on the circle at the corresponding angle. To get a wave out of this continuous rotation, we can look at what happens to just one of the two coordinates this point, which will be given by

$$x(t) = \cos(\theta(t))$$
$$y(t) = \sin(\theta(t)).$$

This process is illustrated by Fig. 1.8.

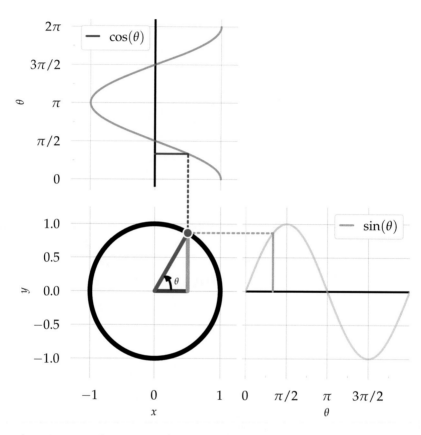

Figure 1.8 A point traveling around the unit circle produces sinusoidal waves when you look only at its horizontal (cosine) or vertical (sine) coordinate determined by its angle θ.

As long as $\theta(t)$ keeps moving at a constant rate, there will be some minimal time t_0 such that $\theta(t_0) = 0$ returns to its initial position: the first complete cycle. We can translate this fundamental period t_0 into a frequency $f_0 = 1/t_0$, measuring the number of cycles completed in one second.

If we assume that $\theta(t)$ is moving at a constant speed, we would have

$$\theta(t) = 2\pi \cdot f_0 \cdot t,$$

so that after $t = t_0$ seconds, $f_0 \cdot t_0 = 1$, and $\theta(t_0) = 2\pi \equiv 0$ because 2π radians completes a turn around the circle. Typically, we don't bother with separate notation for $\theta(t)$. Instead, we parameterize the x and y positions directly in terms of time and frequency:

$$x(t) = \cos\left(2\pi \cdot f_0 \cdot t\right)$$
$$y(t) = \sin\left(2\pi \cdot f_0 \cdot t\right)$$

Wave parameters

The equations above are helpful for providing a geometric interpretation of sinusoids, but they aren't quite general enough to describe every wave we might want. Fortunately, it's not too much more work to do just that.

Every wave can be expressed in the following standardized form:

$$x(t) = A \cdot \cos\left(2\pi \cdot f \cdot t + \phi\right). \tag{1.3}$$

This form requires us to specify three parameters to describe the wave:

- A: the *amplitude* of the wave, how high it can rise or fall from 0;

- f: the *frequency* of the wave, how many cycles it completes in one second;

- ϕ: the *phase* offset of the wave, the starting position (in radians) at time, $t = 0$.

Changing the amplitude A has the effect of stretching or compressing the wave *vertically*. Amplitude (A) can be also be interpreted as the radius of the circle along which our point is traveling.

Changing the frequency f has the effect of stretching or compressing the wave *horizontally*. Frequency (f) can also be thought of as the speed at which our point travels.

Changing the phase offset ϕ has the effect of horizontally *shifting* the wave. Positive ϕ moves the wave to the left, and negative ϕ moves the wave to the right.

Some gotchas   Every wave can be expressed in the above standard form, but that doesn't mean the representation is unique. In general, there are multiple combinations of (A, f, ϕ) that each describe the same wave. For example,

$$x(t) = 2 \cdot \cos(2\pi \cdot f \cdot t) = -2 \cdot \cos(2\pi \cdot f \cdot t - \pi). \tag{1.4}$$

This particular case arises because a phase offset of π amounts to exactly half a turn around the circle. Fig. 1.10 demonstrates this effect: offset by $\phi = -\pi$ turns the wave upside-down, and if we also negate the amplitude (not pictured), the two waves would match up exactly.

Most of the time, this ambiguity of representation doesn't cause us too many troubles. It's sometimes useful to know about this, though, especially when algebraically manipulating wave equations. We won't do too much of that here, but it's good to keep in mind.

1.2.5   Properties of waves

We'll round out this section with a few facts about waves that will be useful in the later chapters.

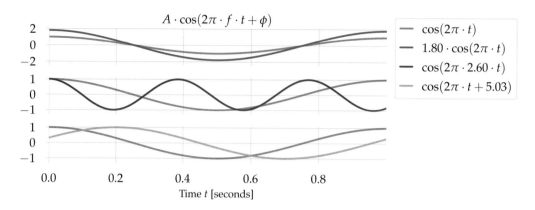

Figure 1.9 Top: changing amplitude (A) stretches or compresses the wave *vertically*. Middle: changing frequency (f) stretches the wave *horizontally*. Bottom: changing phase (ϕ) rigidly *shifts* the wave horizontally with no stretching.

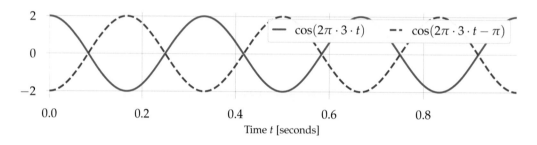

Figure 1.10 A phase offset of $\phi = \pm\pi$ negates any sinusoid.

Symmetry and anti-symmetry

The cosine function has a horizontal symmetry: $\cos(\theta) = \cos(-\theta)$. (Sometimes called an *even function*.) This can be understood geometrically: positive θ means counterclockwise rotation, and negative θ means clockwise rotation. In both cases, the horizontal position is the same as long as the absolute value of the angle is preserved.

By contrast, the sine function as a horizontal *anti-symmetry*: $\sin(-\theta) = -\sin(\theta)$. This also can be understood geometrically as above, except now changing the direction of the rotation means flipping vertically across the horizontal axis.

These symmetries are summarized in Figure 1.11 by observing that negating the angle $-\theta$ preserves horizontal position of the point on the circle but negates the vertical position.

Converting between sine and cosine

Every cosine can be expressed as a sine with an appropriate phase offset, and vice versa. The rule for this is as follows:

$$\sin(\theta) = \cos(\pi/2 - \theta)$$
$$\cos(\theta) = \sin(\pi/2 - \theta)$$

(1.5)

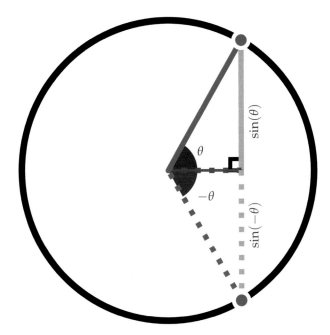

Figure 1.11 Sine and cosine symmetries for θ and $-\theta$. Both angles produce points with the same horizontal component: $\cos(-\theta) = \cos(\theta)$. Opposite angles produce points with opposite vertical components: $\sin(-\theta) = -\sin(\theta)$.

These rules follow from the fact that all triangles have angles which sum to π radians, and that we're dealing specifically with right triangles, as visualized in Figure 1.12. If one angle is θ and the corner has angle $\pi/2$ (90 degrees), then the remaining angle must be $\pi/2 - \theta$, so that the sum totals to π:

$$\theta + \frac{\pi}{2} + \left(\frac{\pi}{2} - \theta\right) = \frac{\pi}{2} + \frac{\pi}{2} + \theta - \theta = \pi.$$

Finally, the cosine of one angle is the sine of the other (and vice versa). In the example below, $\cos(\theta)$ is the ratio of the adjacent side (in red) to the hypotenuse (in blue), which is also the sine of the opposing angle. Similarly, $\sin(\theta)$ is the ratio of the opposite side (in yellow) to the hypotenuse (in blue), which is the cosine of the opposing angle $\pi/2 - \theta$.

Averaging over time

Imagine that we have a wave with some frequency f_0, or equivalently, a fundamental period $t_0 = 1/f_0$. If you were to look at the average value of the wave over the time interval $[0, t_0]$, it will always come out to 0. This may seem like an obvious fact, but it is worth reasoning through carefully.

A careful proof of this statement can be done with calculus, but we can also reason about it intuitively via symmetry, as illustrated in Figure 1.13:

- The time interval $[0, t_0]$ corresponds to one full rotation around the circle.

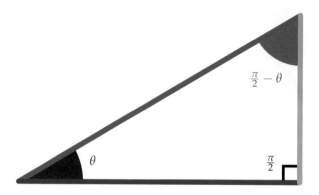

Figure 1.12 If one acute angle θ of a right triangle is given, the other can be inferred as $\pi/2 - \theta$.

- The circle has equally much perimeter in one half of the plane (e.g., $x < 0$) as the other (e.g., $x > 0$).

- For every position (x, y) the point on the circle occupies during our time interval, it must also pass through the opposite point $(-x, -y)$ at some other time during our time interval. These two points cancel each other out and sum to 0.

- Since all points are canceled to 0, so is the total sum.

Note that the above argument only works if the time interval covers exactly one cycle of the wave. The argument can be extended to cover other intervals of the same duration, or intervals spanning a whole number of cycles. However, it **does not** extend to fractional cycles. If you don't cover exactly a whole number of cycles, the average will not be 0. This observation will become important later on when using the Fourier transform to analyze the contents of signals.

1.3 UNITS AND DIMENSIONAL ANALYSIS

Now would be a good time to pause and solidify some concepts.

In the previous section, we learned about sinusoidal waves, which have a standard form

$$x(t) = A \cdot \cos(2\pi \cdot f \cdot t + \phi). \tag{1.6}$$

In code, this would look something like:

```python
import numpy as np


def x(t, amplitude, frequency, phase):
    '''Compute the value of the wave at time t
    with a given amplitude, frequency, and phase offset
    '''
    return amplitude * np.cos(2 * np.pi * frequency * t + phase)
```

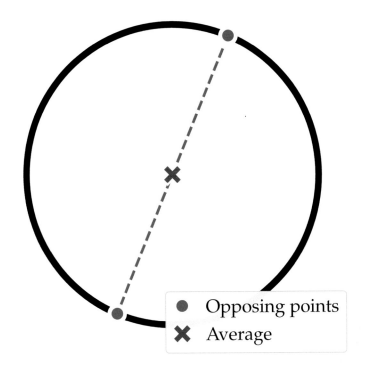

Figure 1.13 Two points at opposite ends of the circle will average to the circle's origin.

For many new-comers (and not-so-new-comers) this string of letters can be difficult to remember and reassemble when necessary. What I want to stress in this function is that this becomes much easier when you can reason about what the pieces mean. To do that properly, we'll need to think about the units of measurement involved. Once we have each variable with its corresponding units, it's easier to see the right way to arrange them, and then you won't have to memorize so much.

1.3.1 Defining our variables

In the form above, our signal $x(t)$ has three fundamental properties:

- A, the amplitude;

- f, the frequency; and

- ϕ, the phase offset.

Now, imagine that we've temporarily forgotten the exact form of (1.6), but we know it involves amplitude, frequency, and phase. How could we reconstruct (1.6)?

Remember from the previous section: cosines (and sines) take angles as their input, and produce a distance measurement (positive or negative) as output. However we combine A, f, and ϕ, we know that whatever input we give to cos must be interpretable as an *angle*.

For now, let's not worry about the amplitude A, since that happens outside the cos (or the call to `np.cos` in the code above). What are each of the remaining quantities measuring, and in what units?

- Time t is measured in [seconds]

- Frequency f is measured in [cycles per second] or [cycles / second]

- 2π is measured in [radians per cycle]

- phase ϕ is measured in [radians]

So if you multiply $2\pi \cdot f \cdot t$, you'll get a product of the form:

$$\left(2\pi \left[\frac{\text{radians}}{\text{cycle}}\right]\right) \cdot \left(f \left[\frac{\text{cycles}}{\text{second}}\right]\right) \cdot (t\,[\text{seconds}])$$

$$= \left(2\pi \left[\frac{\text{radians}}{\cancel{\text{cycle}}}\right]\right) \cdot \left(f \left[\frac{\cancel{\text{cycles}}}{\cancel{\text{second}}}\right]\right) \cdot (t\,[\cancel{\text{seconds}}])$$

$$= 2\pi \cdot f \cdot t\,[\text{radians}]$$

Keeping track of units allows us to see that we were dividing seconds by seconds (which cancel each other), and cycles by cycles (which also cancel). The end result must be in units of radians.

Because ϕ was also in radians, it is okay to add the two together, as is done in the equation for $x(t)$. If one or the other of these was not in units of radians, we could not add them directly, or interpret the answer as an angle. For example, if we tried $2\pi \cdot t/f$, we'd get the following:

Note (This derivation is incorrect! Do not do this).

$$\left(2\pi \left[\frac{\text{radians}}{\text{cycle}}\right]\right) \cdot (t\,[\text{seconds}]) \,/\, \left(f \left[\frac{\text{cycles}}{\text{second}}\right]\right)$$

$$= \left(2\pi \left[\frac{\text{radians}}{\text{cycle}}\right]\right) \cdot (t\,[\text{seconds}]) \cdot \left(\frac{1}{f} \left[\frac{\text{seconds}}{\text{cycle}}\right]\right)$$

$$= 2\pi \cdot \frac{t}{f} \left[\frac{\text{radians} \cdot \text{seconds}^2}{\text{cycle}^2}\right]$$

The result has units of radian-seconds-seconds per cycle per cycle (yikes!), and that is definitely not the same as radians. We would therefore not have been able to compute cos of the result – we would have had an error! This is an example of dimensional analysis helping us: because the units at the end of the calculation do not meet our expectation (radians), we know there must have been a mistake somewhere.

Some readers might remember this kind of reasoning from a high-school physics or chemistry class, and might remember that it's called *dimensional analysis*. It's one of the most useful tools that one can pick up in a science classroom!

1.3.2 Why do this?

This may seem like belaboring the point, but it serves a purpose. When multiplying quantities together, you should be able to track what the corresponding units are. This is often the easiest way to check whether your calculation is correct, and is a good skill to develop.

By far, the most common types of mistakes that people make when first learning to work with signals arise from mismatching quantities, such as adding a frequency to a phase angle, or dividing time by frequency rather than multiplying. These are errors that can be caught directly by checking units.

Note that software doesn't really help us here. Most of the time, we simply write down these quantities as numbers, and the computer has no idea what the numbers mean. We therefore must be careful to document our assumptions carefully, so that we can reason about programs. For example, the program given above would be much better with a bit more explanatory documentation, like so:

```python
import numpy as np

def x(t, amplitude, frequency, phase):
    '''Compute the value of the wave at time t
    with a given amplitude, frequency, and phase offset.

    Parameters
    ----------
    t : time in seconds (real number)

    amplitude : amplitude scaling (real number)

    frequency : cycles per second (real number)

    phase : offset in radians (real number)
    '''
    return amplitude * np.cos(2 * np.pi * frequency * t + phase)
```

Because we're writing in Python, which allows us to define functions without clearly specifying the data types that it supports, we need to be explicit in documenting what the function expects (e.g., **real number**). However, even in strongly typed languages like Java (or, to some extent, C++), the computer still wouldn't understand the *semantics* of the different variables, like that time is measured in seconds (and not milliseconds or fortnights), or that phase is in radians and not degrees.

1.4 AUDIO AND SIGNALS

We'll finish this chapter off with a quick overview of acoustic signals: how waves propagate in an environment, and how we can reason about them as signals.

This introduction is superficial, and will mainly become relevant much later on when we go into convolutional filtering, but it may help to ground some of the principals in physical reality. For a more thorough introduction to acoustics, readers should consult [EP15].

1.4.1 Sound propagation

Let's imagine a simple recording scenario. There is some source emitting sound signal x_s, perhaps a musical instrument or a human voice, and a microphone is placed 5 meters away, recording the microphone signal x_m. To keep things simple, let's imagine that there are no other sounds in the environment, and there are no surfaces that could cause reflections or resonances.

If the speed of sound is some C [meters / second], then it will take $5/C$ [seconds] for sound to travel from the source to the microphone. Put another way, the pressure we measure at time t is induced by the source signal at time $t - 5/C$. More generally, if the distance is D [meters], we have the following

$$x_m(t) \approx x_s(t - D/C). \tag{1.7}$$

Negative time offset?

Many people find the negative time offset in (1.7) to be counter-intuitive, and it's a common mistake to instead write $t + D/C$.

It can help to reason about what $x_m(t)$ represents: it's recording a signal that must have already occurred. If $x_m(t)$ depended on $x_s(t + D/C)$, then it must be recording from the future, which is impossible.

We've written $\approx$ here, and not $=$, because the signals will not be identical: the amplitude will also change. This phenomenon is illustrated in Figure 1.14.

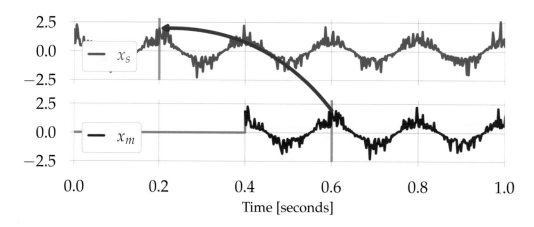

Figure 1.14 A signal x_s emitted by a source (*top*) and observed by a microphone as x_m (bottom). If the signal takes time 0.4 [seconds] to propagate from the source to microphone, then $x_m(t) = x_s(t - 0.4)$.

1.4.2 Pressure, intensity, amplitude, and loudness

The first property of audio signals that we often think of is *loudness*. This turns out to be a surprisingly subtle property: loudness is ultimately a psycho-acoustic phenomenon (something that happens in our brains), and not a physically measurable phenomenon. There are, however, numerous physical quantities that we *can* and do measure that relate to loudness. We'll briefly summarize the most important ones here, as described in the ISO 80000-8:2020 standards:

Name	Description	Symbol	Units
Sound pressure	the change in ambient (static) air pressure due to the sound (see Figure 1.15)	p	Pascals [Pa]
Sound power	the rate at which sound energy is emitted by a source	P	Watts [W]
Sound intensity	sound power per unit area	I	Watts per square-meter $\left[\frac{W}{m^2}\right]$

Physical units

If you're a little out of practice with physical units, **don't worry about it**. We include this information here for completeness, but it mostly won't matter too much when working with digital signals.

For reference though, a Pascal [Pa] is a unit of pressure: force spread over a surface area. If you're from one of the few countries using "standard" units, think of "pounds per square-inch (PSI)" used to measure air pressure in tires: it's the same idea.

Human ears and microphones respond to *sound pressure* (p), the latter converting pressure to an electrical signal measured in Volts [V].

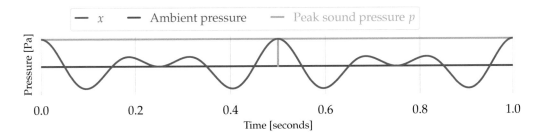

Figure 1.15 An acoustic signal x causes air pressure to fluctuate around the ambient pressure. The amplitude of the signal is given by the maximum difference from the ambient pressure.

Sound power (P) and intensity (I) are useful for quantifying the transmission of sound through an environment. Intensity might seem a bit strange at first glance, as it divides a unit of power (Watts) by a unit of surface area (square-meters). This can be understood by imagining an omni-directional sound source with some fixed power P. As sound propagates from the source, its power is spread over the surface of a sphere. When the sound has traveled r meters from the source, the corresponding sphere has surface area

$$a = 4\pi \cdot r^2.$$

and the corresponding intensity I is

$$I = \frac{P}{a} = \frac{P}{4\pi \cdot r^2} \left[\frac{\text{W}}{\text{m}^2}\right]$$

Without going too far into the details of these quantities, they are defined such that intensity (and power) is proportional to the *square* of pressure:

$$I \propto p^2$$

Equivalently, this implies that sound pressure is inversely proportional to distance:

$$p \propto \sqrt{I} = \sqrt{\frac{P}{4\pi \cdot r^2}} = \frac{1}{r} \cdot \sqrt{\frac{P}{4\pi}} \propto \frac{1}{r}$$

This tells us that if the power P is held constant, and we move a microphone away from the source, the output voltage will decay like $1/r$. In the context of this text, the voltage output of a microphone is not observed directly but is instead post-processed and amplified before being digitized. The resulting *amplitude* (A) of the signal is generally proportional to the microphone's output voltage, which is in turn proportional to pressure:

Tip. Signal amplitude $\propto$ Voltage $\propto$ Sound pressure

---

Proportionality

The proportionality symbol $\propto$ means that two quantities are multiplicatively related to each other, but the exact proportion is not too important (as long as it is constant).

For example, one could write that the circumference of any circle is proportional to its radius:

$$C = 2\pi \cdot r \propto r.$$

if we didn't care too much about the constant factor of 2π. However, we could *not* write $C \propto r^2$, because there is no constant that relates the square of the radius to the circumference.

While this notation does make dimensional analysis impossible, it can be helpful to isolate the key quantities of interest.

Decibels (dB)

Human hearing is sensitive to sound pressure, but this sensitivity is not *linear*. A *doubling* of sound pressure **does not** result in the perception of being *twice as loud*. As mentioned above, perception of loudness is a complex phenomenon, but to a first approximation, it is fair to say that humans are sensitive to *ratios* of pressure (p_1/p_2), not *differences* ($p_1 - p_2$). Logarithms allow us to convert ratios into differences:

> **Tip.** If you're rusty on logarithms, you may want to pause here and read through the *appendix*.

$$\log \frac{a}{b} = \log(a) - \log(b),$$

which are often more convenient to deal with. This leads us to the **decibel (dB)** scale for ratios:

$$\mathrm{dB}(v_1, v_2) = 10 \cdot \log_{10}\left(\frac{v_1}{v_2}\right).$$

Since the logarithm function expects a dimension-less input, the values being compared must have the same units, which cancel when divided.

Sound intensity level is defined in terms of the decibels between sound intensity I and a reference intensity I_0:

$$L = 10 \cdot \log_{10}\left(\frac{I}{I_0}\right),$$

or in terms of pressure (with reference pressure p_0):

$$10 \cdot \log_{10}\left(\frac{p^2}{p_0^2}\right) = 10 \cdot \log_{10}\left(\frac{p}{p_0}\right)^2 = 20 \cdot \log_{10}\left(\frac{p}{p_0}\right).$$

Although we usually think of decibels as measuring intensity (or some nebulous correlate of *loudness*), decibels in fact just measure *ratios*, and can be applied to all kinds of quantities. When specifically applied to sound pressure measurements, we sometimes write $\mathrm{dB_{SPL}}$

Some facts about decibels

- Decibels can be positive or negative. A ratio < 1 corresponds to negative dB; a ratio > 1 corresponds to positive dB. A ratio of exactly 1 gives 0 dB.

- A decibel is one-tenth of a Bel. (But nobody really measures things in Bels.)

- An amplitude (or pressure) ratio of 2 (doubling) corresponds to $20 \log_{10} 2 \approx +6 \, [\mathrm{dB}]$.

- A pressure ratio of $1/2$ (halving) gives $-6 \, [\mathrm{dB}]$.

- The standard reference pressure is $p_0 = 20$ micro-Pascals.

- Wikipedia gives a helpful list of familiar sound pressure levels in decibels.

  - Typical speaking volume is in the range of 40-60 dB.

- The range of typical human hearing (quietest sound to loudest sound) is a complex phenomenon and depends on frequency, but is in the neighborhood of $100 \, [\mathrm{dB}]$.

1.4.3 Frequency and pitch

After intensity, the next most salient feature of audio signals is *frequency* content. By definition, any periodic signal has a fundamental frequency f_0, but most signals are not completely periodic. As we will see much later on, there is a way to reason about the frequency content of arbitrary signals. For now, it will suffice to assume that any signal generally consists of a combination of multiple sinusoidal signals, each one having a distinct fundamental frequency f_0.

Like *loudness* and *amplitude*, it is helpful to distinguish between *pitch* and *frequency*. **Pitch** is a perceptual feature of sound that exists in our brains, but **frequency** is a physical property of sound that can be measured.

> **Note.** Tuning systems In this section, and throughout the text, we will make the simplifying assumption of equal temperament when discussing pitch.

Multiplicative vs. additive

The most important, distinctive aspect of frequency is that it is *multiplicative* rather than *additive*. For example, if we have a note at frequency $A_4 = 440$ [Hz], the next note of the same pitch class is $A_5 = 880 = 440 + 440$. If we again add 440 to A_5, we do not get A_6, but rather something close to $E_6 \approx 1319[\text{Hz}] \approx 3 \times 440 = 1320$. To get A_6, we would have to *double* $A_5 = 880 + 880 = A_4 \cdot 4$.

This observation is perhaps most intuitive to players of fretted string instruments (e.g., guitar), as it explains why the distance between frets gets narrower as one moves up the neck: frequency of a vibrating string is inversely proportional to its length. Players of fret-less stringed instruments also learn this, but frets make it much easier.

More generally, the size of an octave (measured in Hz) is *not constant*. Instead, the size doubles each time we go up by an octave. Similarly, the size of the octave B_4–B_5 is not the same as the octave A_4–A_5. However, the *ratio* of frequencies separated by an octave is always constant (2) regardless of which octave or specific frequency we're talking about.

One way to think about this multiplicative behavior from some f to $f_2 = 2 \cdot f$ is that f_2 has 2 cycles for every 1 cycle in f, as depicted below depicted in Figure 1.16.

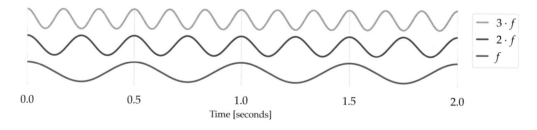

Figure 1.16 Waves at frequencies $f = 2$ [Hz], $2 \cdot f$, and $3 \cdot f$. Note that $2 \cdot f$ and $3 \cdot f$ complete whole cycles for each cycle of f, but not for each other.

The multiplicative nature of frequency means that we need to consider ratios rather than differences when comparing frequencies. The octave relation $(f, 2 \cdot f)$ gives a starting point for generating tuning systems, which map frequencies onto pitches and pitch-classes.

We could spend quite some time on tuning systems, but it's a bit beyond the scope of what we'll need. For better or worse, we'll focus on 12-tone equal temperament (12-TET), which divides each octave "evenly" into 12 frequencies (*semitones*). The notion of "evenly" is that the ratio of each successive pair of frequencies remains constant, and after stepping through 12 of them, we have a full octave (ratio of 2). This implies that each ratio must be $2^{1/12} \approx 1.06$.

While semitone divisions of the octave are useful in many musical contexts, they may be too coarse for other applications where we could benefit from a more fine-grained representation of frequency. This motivates the use of **cents** (denoted ¢), which divide each semitone range evenly into 100 pieces; or, equivalently, each octave into 1200 pieces. A change in frequency of 1¢ therefore corresponds to a multiplicative factor of $2^{1/1200} \approx 1.0006$.

Frequency ranges

Finally, to ground the notion of frequency in reality, Figure 1.17 illustrates familiar frequency ranges.

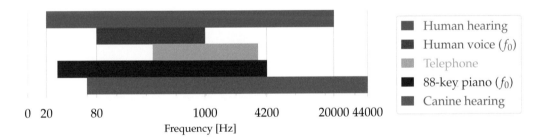

Figure 1.17 A visual comparison of important (and not-so-important) frequency ranges.

1.5 EXERCISES

Exercise 1.1. Write an equation describing a continuous sinusoid $x(t)$ oscillating at 5 cycles per second.

Exercise 1.2. For each of the following waves, convert them into an equivalent expression in standard form (as in (1.3)). What is the amplitude (A), frequency (f) and initial phase (ϕ) for each one?

1. $x(t) = \sin(2\pi \cdot 3 \cdot t)$

2. $x(t) = 2 \cdot \cos\left(\pi \cdot t - \frac{\pi}{2}\right)$

3. $x(t) = -\sin(4\pi \cdot t + \pi)$

Exercise 1.3. A waveform is described to you in terms of non-standard units:

- $f = 600 \left[\frac{\text{half-cycles}}{\text{minute}} \right]$

- $\phi = 120°$

Convert this into a standard form with f expressed in Hz and ϕ in radians. Use dimensional analysis to check your conversion.

Exercise 1.4. Imagine we have a sound source of constant power, and a microphone at distance r observes sound pressure p.

1. At what distance should we place the microphone to observe sound pressure $p/2$?

2. If the sound pressure level is L [dB] at distance r, what distance would produce a pressure level of $L - 5$ [dB]?

Exercise 1.5. If we assume the average human can hear frequencies between 20 and 20,000 Hz, approximately how many octaves is this range?

## CHAPTER 2

Digital sampling

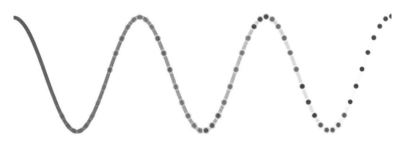

In the previous chapter, signals were assumed to exist in continuous time. A signal was represented as $x(t)$, where t could be any continuous value representing time: $0, 3, -1/12, \sqrt{2}, \dots$ are all valid values for t.

Digital computers, the kind we write programs for, cannot work directly with continuous signals. When a signal enters a digital computer, it must be *discretized* before the computer can be processed. There are two ingredients to discretization: sampling, and quantization.

If you've ever looked at the file information for digital audio on a computer, you may have seen something like the following (generated by the soxi program):

```
$ soxi my_signal.wav

Input File     : 'my_signal.wav'
Channels       : 1
Sample Rate    : 22050
Precision      : 16-bit
```

The `channels` field tells you how many signals are contained in the file: in this case, only 1 because it's monophonic. The other two fields, `sample rate` and `precision` have to do with sampling and quantization.

This section introduces discrete sampling: the process of summarizing a continuous signal $x(t)$ by a discrete sequence of sample values. There is a vast and deep literature around the topic of sampling, and this chapter generally covers the basics.

At the end of the chapter, we'll come back to quantization, but sampling is by far the more important part of the discretization process.

DOI: 10.1201/9781003264859-2

2.1 SAMPLING PERIOD AND RATE

The basic scheme for digitizing an analog signal is to measure the signal $x(t)$ at a sequence of uniformly spaced time points. The **sampling period** is denoted by t_s, which must be a positive number, measuring the number of seconds between samples. (Fractional values are allowed.) The resulting sequence of samples will be

$$x(0), x(0 + t_s), x(0 + t_s + t_s), \ldots$$

More generally, the nth sample (for an arbitrary integer $n = 0, 1, 2, \ldots$) represents the signal at time $n \cdot t_s$. In a slight abuse of notation, we use square brackets with index n to indicate discrete signals $x[n]$, and parentheses with time t to denote continuous signals:

$$x[n] = x(n \cdot t_s). \tag{2.1}$$

By convention, we use N to denote the total number of samples.

Oftentimes, it is more convenient to work with the **sampling rate**, which we denote as

$$f_s = \frac{1}{t_s} \quad \left[\frac{\text{samples}}{\text{second}}\right].$$

Note that the sampling rate can always be converted to a sampling period (and vice versa) by taking reciprocals, resulting in the following (equivalent) form for discretely sampled signals:

$$x[n] = x\left(\frac{n}{f_s}\right). \tag{2.2}$$

2.1.1 Discrete signals and visualization

A signal $x[n]$ which has been sampled is said to be a *discrete-time* signal (or sometimes, just *discrete signal*), to distinguish it from *continuous-time* signals $x(t)$.

Properly speaking, we do not have sample values at non-integer indices (e.g., $x[n + 1/2]$), but it is often helpful for understanding to connect sample values in visualizations. When visualizing discrete-time signals, we adopt the convention of using *step-plots* rather than continuously varying curves (like $x(t)$ in Figure 2.1), to emphasize the fact that the signal has been discretized. Step plots, as demonstrated in Figure 2.2, preserve the sample value $x[n]$ up to the next sample position $x[n + 1]$.

For most practical applications, we tend to have sampling rates that are sufficiently high to guarantee that a step plot and a continuous plot look visually identical. However, throughout this text, we will often use examples generated at low sampling rates because they are easier to understand.

2.1.2 Tone generation

We now have everything that we need to start making sounds. In this example, we'll generate a pure tone at 220 Hz.

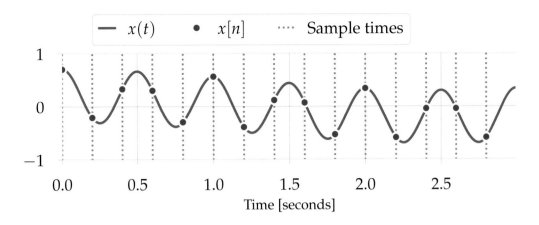

Figure 2.1 A continuous signal (solid curve) sampled at $t_s = 0.2$ [seconds/sample] (dots).

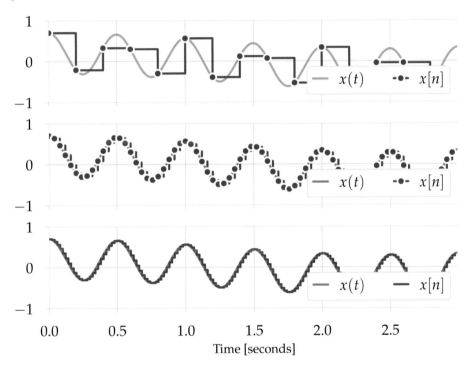

Figure 2.2 A continuous signal $x(t)$ and the discrete signal $x[n]$ obtained by sampling at 5 Hz (top), 25 Hz (middle), and 50 Hz (bottom). The discrete-time signal is illustrated as a step plot. At sufficiently high sampling rates, the continuous and discrete signals appear visually similar.

Recall from the previous chapter that a wave at frequency f_0 is expressed as a function of time t by

$$x(t) = \cos\left(2\pi \cdot f_0 \cdot t\right).$$

(Here, we'll ignore amplitude and phase to keep things simple.)

Even if we don't have an existing signal to sample, we can still sample from this idealized signal by substituting $t = n/f_s$ and computing the cosine values directly:

$$x[n] = \cos\left(2\pi \cdot f_0 \cdot \frac{n}{f_s}\right),$$

or, in code:

```
for n in range(N):   # n goes from 0, 1, 2, ..., N-1
    x[n] = np.cos(2 * np.pi * f_0 * n / f_s)
```

Vectorization

In practice, generating a tone in this fashion would be rather slow – at least, when using the Python programming language. Instead, a much faster way to do it is to pre-allocate all the values of **n** as a *vector*

```
n = [0, 1, 2, ..., N-1].
```

> **Note.** For our purposes, the word *vector* just means a list or array of numbers.

In code, you can do this by using the `np.arange` function, like so:

```
n = np.arange(N)   # n is now an array containing values [0, 1, 2, ..
↪., N-1]
x = np.cos(2 * np.pi * f_0 * n / f_s)
```

The result will be an array **x** with **N** total samples. Python (**numpy**) is smart enough to know that when you multiply, add, and call cosine on lists of numbers, it should apply these operations to each element of the list. This process is called *vectorization*, and it's quite common in numerical computing. It might look a little strange and take some getting used to, but it does simplify and accelerate many of the types of operations we do in signal processing code.

```
# We use numpy for numeric computation
import numpy as np

# The Audio object allows us to play back sound in Jupyter
from IPython.display import Audio

# We'll use an 8 KHz sampling rate, roughly
# equivalent to telephone quality
fs = 8000

duration = 1.5   # And generate 1.5 seconds of audio
```

(continues on next page)

(continued from previous page)

```
# Total number of samples, use int(...) to round to whole number
N = int(duration * fs)

f0 = 220   # Generate a pure tone at 220 Hz

n = np.arange(N)   # Make an array of sample indices

# And make the tone, using (n / fs) in place of t
x = np.cos(2 * np.pi * f0 * n / fs)

# How does it sound?
Audio(data=x, rate=fs)
```

2.2   ALIASING

The previous section introduced uniform sampling, which allows us to represent a continuous signal $x(t)$ by a discrete sequence of sample values $x[n]$.

In this section, we'll see that this idea comes with some restrictions.

2.2.1   What is aliasing?

Aliasing is the name we give to the phenomenon when two distinct continuous signals $x_1(t)$ and $x_2(t)$ produce the same sequence of sample values $x[n]$ when sampled at a fixed rate f_s. More specifically, we usually think of aliasing in terms of pure (sinusoidal) tones $x(t) = A \cdot \cos\left(2\pi \cdot f \cdot t + \phi\right)$.

Theorem 2.1 (Aliasing). Given a sampling rate f_s, two frequencies f and f' are **aliases** of each other if for some integer k,

$$f' = f + k \cdot f_s. \tag{2.3}$$

If sampled at rate f_s, two waves x (at frequency f) and y (at frequency f')

$$x[n] = A \cdot \cos\left(2\pi \cdot f \cdot \frac{n}{f_s} + \phi\right)$$
$$y[n] = A \cdot \cos\left(2\pi \cdot f' \cdot \frac{n}{f_s} + \phi\right)$$

will produce identical samples: $x[n] = y[n]$ for all $n = 0, 1, 2\ldots$.

Or, in words, frequency f' is f plus some whole number multiples of the sampling rate f_s. Equation (2.3) is known as the *aliasing equation*, and it tells us how to find all aliasing frequencies for a given f and sampling rate.

Fig. 2.3 illustrates this effect: for any f that we choose, once a sampling rate f_s is chosen, there are infinitely many frequencies that produce an identical sequence of samples.

> **Note (Aside: aliasing units).** The aliasing equation combines a frequency f [cycles/sec] with a sampling rate f_s [samples/sec], scaled by an integer k. Following our earlier discussion of *dimension analysis*, we'll need to assign some units to k for this to be well-defined. Since f' should also be in [cycles/sec], we should define k as [cycles/sample] so that $k \cdot f_s$ has units [cycles/sec] and can be added to f.
>
> This is not just being pedantic: giving units to k helps us reason about what it means intuitively. It measures additional **cycles between samples**.

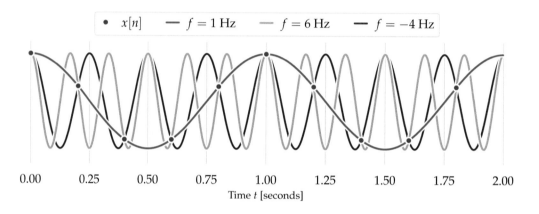

Figure 2.3 An example of aliasing frequencies for $f = 1$ Hz when the sampling rate is $f_s = 5$ Hz.

2.2.2 Why is aliasing a problem?

Aliasing is an unavoidable consequence of digital sampling: there will always be frequencies that look the same after sampling. The consequence of this fact is that once you've sampled a signal, you may not be able to determine the frequency of the wave that produced the samples you've observed.

We'll see in the next section that the Nyquist-Shannon theorem suggests a resolution to this problem, but for now, let's work toward a better understanding of why aliasing occurs.

> **Negative frequency?**
>
> You may have noticed that we can have $k < 0$ in the aliasing equation, which can produce frequencies like $f = -4$ Hz. What does this mean?
>
> You can think of "negative frequency" as the frequency of a point traveling clockwise around the circle, rather than counter-clockwise. However, if you look carefully, you'll notice that the curve for $f = -4$ looks just like a wave at $f = +4$, at least for cosine waves. (For sine waves, the picture is almost the same, but one of the waves would have to flip upside-down.)
>
> We'll see in the next section why this idea is so important.

2.2.3 Why does aliasing happen?

In the previous section, we sampled a pure tone by using the following equation

$$x[n] = \cos\left(2\pi \cdot f \cdot \frac{n}{f_s}\right).$$

To see why aliasing occurs between frequencies f and $f + k \cdot f_s$, we can plug the latter into the equation above and see what happens. The key idea that we'll need is the following identity:

$$\cos(\theta + 2\pi \cdot m) = \cos(\theta) \quad \text{for any integer } m. \tag{2.4}$$

This identity works because if m is an integer, then $2\pi \cdot m$ is a whole number of rotations around the circle (clockwise if $m < 0$, counter-clockwise if $m > 0$). As a result, adding $2\pi \cdot m$ to any angle θ leaves you back at θ.

By analogy, you can think of θ as the minute hand on a clock, and each m counts an hour offset. Regardless of the current time, adding one hour (or two, or three, counted by m) will leave the minute hand in exactly the same place.

Proof of aliasing

The proof of *Theorem 2.1* uses a bit of algebra to re-arrange the terms of the equation, and then exploits the identity defined above to simplify the equation. We start with the definition for the sequence generated by sampling a pure tone at $f + k \cdot f_s$, and work our way toward the sequence generated by f. We'll ignore amplitude A and phase offset ϕ to avoid cluttering the notation, but the argument goes through just as well when those features are included.

Proof.

$$\cos\left(2\pi \cdot (f + k \cdot f_s) \cdot \frac{n}{f_s}\right) = \cos\left(2\pi \cdot f \cdot \frac{n}{f_s} + 2\pi \cdot k \cdot f_s \cdot \frac{n}{f_s}\right) \quad \begin{array}{l}\text{Distribute} \\ \text{multiplication}\end{array}$$

$$= \cos\left(2\pi \cdot f \cdot \frac{n}{f_s} + 2\pi \cdot k \cdot \cancel{f_s} \cdot \frac{n}{\cancel{f_s}}\right) \quad \text{Cancel } \frac{f_s}{f_s} = 1$$

$$= \cos\left(2\pi \cdot f \cdot \frac{n}{f_s} + 2\pi \cdot k \cdot n\right) \quad k \cdot n \text{ is an integer}$$

$$= \cos\left(2\pi \cdot f \cdot \frac{n}{f_s} + \cancel{2\pi \cdot k \cdot n}\right) \quad \begin{array}{l}\text{Cancel extra whole} \\ \text{cycles}\end{array}$$

$$= \cos\left(2\pi \cdot f \cdot \frac{n}{f_s}\right)$$

$$= x[n].$$

$\square$

This shows that the two frequencies f and $f + k \cdot f_s$ produce the same sequence of samples, regardless of k. (But definitely depends on f_s.)

If we were to listen to sampled tones at these frequencies, we shouldn't be able to tell them apart. Let's test that hypothesis.

2.2.4 Example: aliased tones

The following code example generates two tones at aliasing frequencies but is otherwise similar to the example in the previous section. The two sequences of samples will be numerically identical, and therefore sound identical.

Here, we just used $k = 1$, but any integer k will produce the same results.

```python
import numpy as np
from IPython.display import display, Audio

fs = 8000   # Sampling at 8 KHz

f0_original = 220   # A pure tone at 220 Hz
f0_aliased = f0_original + fs   # An alias at 220 + 8000 = 8220

N = int(duration * fs)   # How many samples is 2 seconds?

n = np.arange(N)   # Generate the sample positions

# Construct the signals
x_original = np.cos(2 * np.pi * f0_original * n / fs)
x_aliased = np.cos(2 * np.pi * f0_aliased * n / fs)

# Let's hear them both
print("fs = {} Hz, f0 = {} Hz".format(fs, f0_original))
display(Audio(data=x_original, rate=fs))

print("fs = {} Hz, f0 aliased = {} Hz".format(fs, f0_aliased))
display(Audio(data=x_aliased, rate=fs))
```

2.2.5 Summary

Although we usually view aliasing as a problem to be overcome, it is not always a bad thing. In some applications (though usually not audio), aliasing can be exploited to efficiently sample high-frequency signals with low sampling rates.

Within the context of audio, one of the most important applications of aliasing is in the development of *Fast Fourier Transform (FFT)* algorithms, covered later in this text.

The important thing is not necessarily to prevent aliasing, but to understand its effects thoroughly.

2.3 THE NYQUIST-SHANNON SAMPLING THEOREM

In the previous section, we saw that aliasing occurs between frequencies related by an integer multiple of the sampling rate f_s:

$$f' = f + k \cdot f_s.$$

The bad news is that we can never avoid this: it's a byproduct of representing continuous signals by discrete samples.

   The good news it that if we're careful, we can ensure that aliasing effects do not corrupt our signals (or analysis). This is what the Nyquist-Shannon theorem is all about: establishing the conditions under which sampling is okay.

Aside: what's a 'theorem'?

   Readers who don't spend a lot of time with mathematical terminology are often confused by the word "theorem", and think it's equivalent to the word "theory". **It is not.**

   A "theorem" is a mathematical fact that logically follows from a set of assumptions.

   Some other vocabulary for mathematical facts that you might not be familiar with:

- **lemma**: a fact used to help prove a larger theorem.

- **corollary**: a fact that follows immediately from a theorem.

Proving that (continuous) signals can be expressed as a combination of sinuoids is a bit out of scope for us here, but we will show in a later chapter that this holds in the discrete case: every discrete signal $x[n]$ can be expressed as a weighted sum of discrete (sampled) sinusoids.

2.3.1 Sampling pure tones and combinations

The sampling theorem is most easily understood in terms of pure tones (sinusoids). While most any signal you encounter out in the world is unlikely to be a pure sinusoid, it turns out that under mild conditions, **every** continuous signal can be expressed as a **combination of sinusoids**. By **combination**, we specifically mean a weighted sum, possibly with different phase offsets for each frequency:

$$\begin{aligned} x(t) =& A_1 \cdot \cos(2\pi \cdot f_1 \cdot t + \phi_1) + \\ & A_2 \cdot \cos(2\pi \cdot f_2 \cdot t + \phi_2) + \\ & A_3 \cdot \cos(2\pi \cdot f_3 \cdot t + \phi_3) + \cdots \end{aligned}$$

When we sample a signal to produce $x[n]$, we can equivalently think of sampling each sinusoid first and then summing the results:

$$
\begin{aligned}
x[n] =& A_1 \cdot \cos\left(2\pi \cdot f_1 \cdot \frac{n}{f_s} + \phi_1\right) + \\
& A_2 \cdot \cos\left(2\pi \cdot f_2 \cdot \frac{n}{f_s} + \phi_2\right) + \\
& A_3 \cdot \cos\left(2\pi \cdot f_3 \cdot \frac{n}{f_s} + \phi_3\right) + \cdots
\end{aligned}
$$

Reasoning about sampling in this way will simplify things quite a bit. If we can understand what sampling does for pure sinusoids, then we can extend that knowledge to general signals. Note that we don't need to know the specific values for A_i or ϕ_i; these quantities will generally be unknown. It suffices to know that they exist.

2.3.2 Band-limited sampling

A signal $x(t)$ is **band-limited** if it can be expressed as a combination (weighted sum) of pure sinusoids whose frequencies lie between some minimum frequency f_- and some maximum frequency $f_+ \geq f_-$. The size of this *band* of frequencies,

$$
f_+ - f_-
$$

is known as the **bandwidth** of the signal.

Another way to think of band-limiting is that any sinusoid with frequency $f < f_-$ or $f > f_+$ has no weight in the combination that produces $x(t)$.

The basic idea of the Nyquist-Shannon theorem is that if the sampling rate f_s is sufficiently large (compared to the bandwidth of the signal), then aliasing can't hurt us: **aliases must have zero amplitude.**

2.3.3 The Nyquist-Shannon sampling theorem

We can now formally state the sampling theorem, commonly attributed to Harry Nyquist and Claude Shannon [Nyq28, Sha49].

> **Attribution**
>
> Although the sampling theorem is named for Nyquist and Shannon, it was also previously discovered by E.T. Whittaker in 1915 [Whi15] and Vladimir Kotelnikov in 1933 [Kot33] (among others).
>
> This is an unfortunate example of Stigler's Law: discoveries are not always named for the people who make them.

We'll actually state a simpler form of their theorem that's sufficient for our needs.

Theorem 2.2 (Nyquist-Shannon). If $x(t)$ band-limited to the range $f_- \ldots f_+$, then any sampling rate $f_s \geq f_+ - f_-$ is sufficient to prevent aliasing.

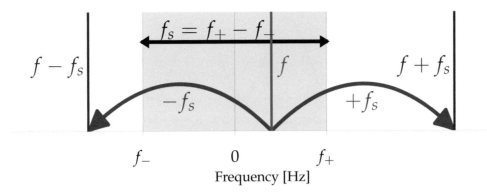

Figure 2.4 The shaded region indicates frequencies within the band limits of the signal. If the sampling rate f_s is sufficiently high, then aliases of f inside the band limits must land outside.

Proof. Pick any frequency f, which will have aliasing frequencies of the form $f' = f + k \cdot f_s$ for integer values k. Because the space between aliases is at least f_s, and the bandwidth of the signal is at most f_s, any aliasing frequency f' must reside outside the frequency range of $x(t)$ as depicted in Fig. 2.4.

As a result, the discrete sampled signal $x[n]$ will depend only on those frequencies within the band limits $f_- < f < f_+$, which cannot be aliases of each other. □

2.3.4 Band-limiting in practice

The Nyquist-Shannon theorem tells us how to choose a sampling rate, provided we know the band limits of the signal(s) we'd like to sample. But how do we ensure that $x(t)$ is actually band-limited?

In hardware analog-to-digital converters (ADCs), this is done by using an analog circuit to filter the continuous signal and remove any frequencies above f_+ prior to sampling. If you've ever seen a tone knob on an electric guitar, the principle is much the same.

Setting the bandwidth

At first glance, the Nyquist-Shannon theorem might suggest to set $f_- = 0$ and f_+ to some reasonable maximum frequency, e.g. for audio, the upper range of human hearing (about 20,000 Hz). Unfortunately, this approach won't work.

To see why, consider the two signals plotted in Figure 2.5: one has a frequency of 5 Hz, and the other has a frequency of -5 Hz:

$$x_1(t) = \cos\left(2\pi \cdot 5 \cdot t\right)$$
$$x_2(t) = \cos\left(2\pi \cdot (-5) \cdot t\right).$$

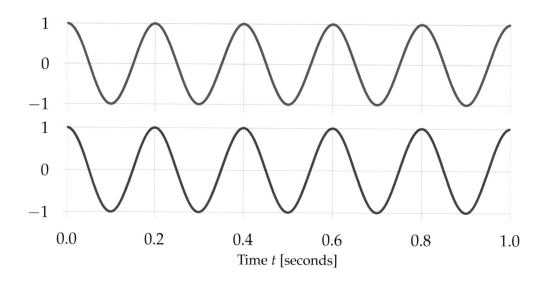

Figure 2.5 Two cosine waves, one with positive frequency and one with negative frequency.

Note (Can you tell which is which?). No. These two frequencies produce identical signals. In general, any wave with a negative frequency can be equivalently represented as a wave with positive frequency and a different phase.

> The situation with negative frequency gets a bit trickier when signals can be complex-valued. In that case, negative frequency *can* be distinguished from positive frequency, in exactly the same way that clockwise and counter-clockwise rotation can be distinguished from each other.
>
> However, we'll only be dealing with real-valued signals here.

For cosine waves, this is just the *symmetry property* from before:

$$\cos(-\theta) = \cos(\theta),$$

which implies

$$\cos\left(2\pi \cdot (-f) \cdot t + \phi\right) = \cos\left(2\pi \cdot f \cdot t - \phi\right).$$

For sine waves, there is an anti-symmetry property, which we can combine with the phase inversion property (eq. (1.4)):

$$\sin(-\theta) = -\sin(\theta) = \sin(\theta - \pi),$$

which implies

$$\sin\left(2\pi \cdot (-f) \cdot t + \phi\right) = \sin\left(2\pi \cdot f \cdot t + \pi - \phi\right).$$

If we wanted to filter out negative frequencies (i.e., set $f_- = 0$), then we must necessarily also filter out positive frequencies as well, because f and $-f$ are indistinguishable from each other. Put another way, for any frequency f that we want to keep, we must also keep $-f$, so our band limits must be symmetric around 0.

Putting it all together

The symmetry argument above tells us that we must have $f_- = -f_+$. This leads to the more common formula for the sampling rate in the Nyquist-Shannon theorem:

$$f_s \geq 2 \cdot f_+, \tag{2.5}$$

because $f_+ - f_- = f_+ - (-f_+) = 2 \cdot f_+$.

Alternatively, for a fixed sampling rate f_s, the highest frequency that can be measured without aliasing artifacts is $f_s/2$, also known as the **Nyquist frequency** (for sampling rate f_s).

For audio applications, we typically want f_+ to be sufficiently large to capture the audible range, which for humans, generally spans 30 to 20000 Hz. This suggests a sampling rate $f_s \geq 2 \cdot f_+ \approx 40000$. Combining this with a few various technological constraints resulted in the standard rate $f_s = 44100$ Hz for compact disc quality audio, and which is still commonly used today.

2.4   QUANTIZATION

If you recall the introduction to this chapter, we saw that digitized audio has two properties: the *sampling rate* (already covered in this chapter), and the *precision*. This section concerns the *precision* of digital audio, but what exactly does this mean? To understand this, we'll first need to take a detour to see how computers represent numerical data.

2.4.1   Background: digital computers and integers

These days, most humans use the Hindu-Arabic (or *decimal*) numeral system to represent numbers. With decimal digits, we use the ten symbols $0, 1, 2, \cdots, 9$ to encode numbers as combinations of powers of ten (the *base* or *radix* of the system). For example, a number like 132 can be expanded out in terms of powers of 10:

$$132 = 1 \cdot 10^2 + 3 \cdot 10^1 + 2 \cdot 10^0.$$

There's nothing magical about the number 10 here: it was probably chosen to match up with the number of fingers (ahem, *digits*) most people possess. Any other *base* can work too.

Of course, computers don't have fingers, so they might find decimal to be difficult. Computers do have logic gates though, which can represent *true* and *false* values, which we can interpret as 1 and 0, respectively. This leads us to *binary* numbers, which only use two symbols to encode numbers as combinations of powers of 2, rather than combinations of powers of 10.

In our example above, the number 132 could be represented as

$$132 = 1 \cdot 128 + 0 \cdot 64 + 0 \cdot 32 + 0 \cdot 16 + 0 \cdot 8 + 1 \cdot 4 + 0 \cdot 2 + 0 \cdot 1$$
$$= 1 \cdot 2^7 + 0 \cdot 2^6 + 0 \cdot 2^5 + 0 \cdot 2^4 + 0 \cdot 2^3 + 1 \cdot 2^2 + 0 \cdot 2^1 + 0 \cdot 2^0,$$

or, more compactly, as 10000100_2 (where the subscript lets us know we're in binary). We refer to each position as a **bit** (short for **b**inary dig**it**).

> **Tip.** Binary numbers in Python Sometimes it's handy to encode binary numbers directly into Python code. You can do this by prefixing the bits with 0b. For example:
>
> ```
> # x = 132
> x = 0b10000100
> ```

For various technical reasons, computers don't generally support arbitrarily large numbers. Instead, integers come in a few different "sizes" depending on how many bits we'll need: usually, 8, 16, 32, or 64 bits. The example above is an *8-bit number*, but it could just as easily have been written in 16-, 32- or 64-bit representation by using leading zeros: 0000000010000100_2 for 132 in 16-bit form.

Negative numbers

An n-bit number can represent 2^n distinct numbers, but *which numbers?* We can interpret the bit representation as the numbers $0, 1, \cdots, 2^n - 1$, but this doesn't provide an obvious way to represent *negative numbers*.

There's an elegant solution to this problem if we imagine arranging binary numbers around a circle, as illustrated in figure 2.6 for 3-bit numbers. We can think of counter-clockwise movement as *incrementing* by one, and clockwise movement as *decrementing* by one. In this view, the numbers beginning with 1 can be seen as negative numbers: $111 = -1, 110 = -2, \ldots$, and the numbers beginning with 0 are the non-negative numbers as discussed above. It's beyond our scope here, but this representation of integers, known as two's complement, has many nice properties, and is implemented by almost every modern computer for doing integer arithmetic.

To summarize, an n-bit two's-complement integer can represent 2^{n-1} distinct non-negative numbers $(0, 1, \ldots, 2^{n-1} - 1)$, and 2^{n-1} distinct negative numbers $(-1, -2, \ldots, -2^{n-1})$. For example, an 8-bit number can take $2^8 = 256$ distinct values: $-128, -127, \ldots, -2, -1, 0, 1, 2, \ldots, 127$.

This is a relatively minor detail in the bigger picture of digital signals, but it can help to understand why quantized signals look the way they do (as illustrated below).

2.4.2 Defining precision and quantization

Precision, also known as **bit depth**, refers to how many *bits* are used to represent each sample in a digital signal. While we typically think of signals as taking on continuous real values, computers **quantize** these values to be drawn from a fixed, finite set of numbers.

High precision means that we have more distinct values and can, therefore, faithfully represent smaller differences and get a more accurate representation of the underlying continuous signal. However, doing so comes at a cost: higher precision means we're storing and transmitting more data. There's a trade-off to be made between storage cost and the perceptual fidelity of the quantized signal. Investigating this thoroughly is beyond the scope of this text, but interested readers are encouraged to look into *perceptual coding* and *lossy audio compression* to learn more.

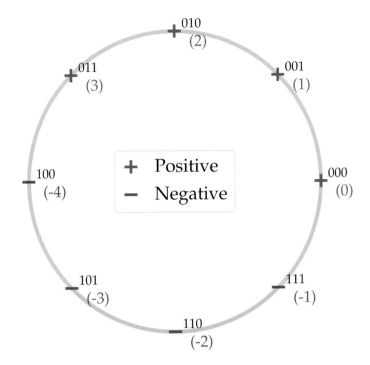

Figure 2.6 Two's complement representation of 3-bit integers (-4, -3, ..., 3).

2.4.3 The effects of quantization

Fig. 2.7 illustrates the effects of varying levels of quantization on samples from a continuous waveform. High precision values (16-bit) provide a good approximation to the original wave, but this approximation deteriorates as we reduce the precision. In the extreme case (1-bit quantization), each sample can take only one of two values (-1 or 0), which results in a highly distorted signal.

2.4.4 Dynamic range

Fundamentally, quantization reduces the number of distinct values that can be observed in a signal. Rather than the full range of continuous values, say voltages in the range $[-V, +V]$, we instead divide up the range into pieces of constant (quantized) value. For uniform quantization into a n-bit integer representation, we have 2^n distinct numbers representing different values between $-V$ and $+V$. If q represents an n-bit quantized integer value $-2^{n-1} \leq q \leq 2^{n-1} - 1$, then we can map q to a value $v(q)$:

$$v(q) = V \cdot \left(\frac{q}{2^{n-1}} \right), \tag{2.6}$$

as illustrated in Fig. 2.7. This process is lossy, in that not every continuous input value can be realized by a specific quantized number q, and this fact raises many questions about how accurate the quantization process is.

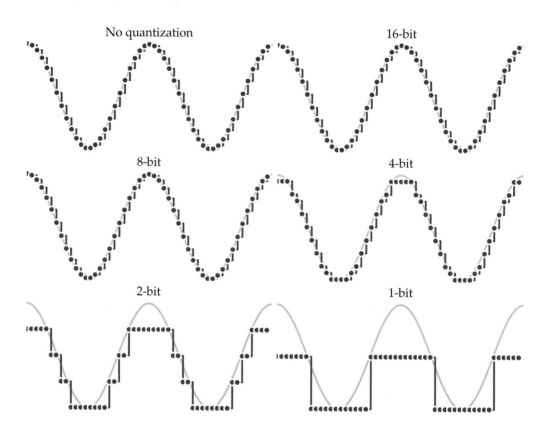

Figure 2.7 A continuous signal (solid curve) is sampled (dots) and then quantized to several different bit depths. At low depths, the sample values are noticeably different from the original signal.

We won't go into all the details of analyzing quantization error, but one commonly used method to evaluate a quantization scheme is to measure its **dynamic range**: the ratio of its loudest value v_+ to its quietest (non-zero) value v_-, measured in decibels. Typically, the values in question are voltages, which you may recall from *chapter 1* are proportional to sound pressure, and that intensity is proportional to the *square* of pressure (and, therefore, the square of voltage). As a result, to compute dynamic range R_{dB} for quantized voltages, we'll need to square them:

$$R_{\text{dB}} = 10 \cdot \log_{10} \left(\frac{v_+}{v_-} \right)^2 = 20 \cdot \log_{10} \frac{v_+}{v_-}. \tag{2.7}$$

To use this idea, we'll need to calculate the smallest and largest (absolute) values attainable by (2.6). The smallest value, v_- is attained by $q = 1$, so that

$$v_- = v(1) = V \cdot \left(\frac{1}{2^{n-1}} \right).$$

The largest absolute value, v_+, is attained by $q = -2^{n-1}$:

$$v_+ = \left| v\left(-2^{n-1} \right) \right| = V \cdot \left(\frac{2^{n-1}}{2^{n-1}} \right).$$

When we form the ratio v_+/v_-, note that there is a common factor of $V/2^{n-1}$ that can be canceled, resulting in:

$$\frac{v_+}{v_-} = 2^{n-1}.$$

Plugging this value into (2.7) yields

$$\begin{aligned} R_{\mathrm{dB}} &= 20 \cdot \log_{10} 2^{n-1} \\ &= (n-1) \cdot 20 \log_{10} 2 \\ &\approx (n-1) \cdot 6.02 \ [\mathrm{dB}] \end{aligned} \tag{2.8}$$

Tip. Equation (2.8) gives rise to a commonly used rule of thumb: each bit of precision adds about 6 dB of dynamic range.

> **Warning.** Note that dynamic range is not to be confused with *signal-to-noise ratio (SNR)*, which is also often reported as a measure of quantization quality, but a bit more involved to calculate.

Equation (2.8) allows us to measure the dynamic range purely as a function of bit depth. The following table gives dynamic range values for several common choices of n.

Precision [bits]	Dynamic range [dB]
$n = 1$	0.0
$n = 2$	6.02
$n = 4$	18.06
$n = 8$	42.14
$n = 16$ (compact disc standard)	90.31
$n = 24$	138.47
$n = 32$	186.64

2.4.5 Floating point numbers

The integer quantization scheme described above is commonly used for storing audio signals (e.g., in `.wav` format), but it is sometimes not the most convenient representation for *processing* audio signals. Instead, most computers use **floating point** representations for numerical data and have dedicated hardware for carrying out mathematical operations on floating point numbers. Floating point numbers differ from integer representations in several key ways, but they are the standard choice for representing fractional numbers and (approximately) continuous values. This is achieved by using a *non-uniform* spacing of numbers encoded in a format similar to scientific notation. Note that floating point numbers are still technically quantized, but the quantization level is practically so small that we treat them as though they were not quantized at all.

Floating point representations are defined by the IEEE 754 standard, which is quite a bit more detailed than we need to get into here. Rather than calculating the minimum and maximum values from the specification – which is doable, but tedious – we'll instead see how to retrieve this information from the computer directly.

Like most numerical computation libraries, Numpy provides an interface for getting information about the number of representations used behind the scenes. For floating point numbers, this is provided by the `np.finfo` function. This function accepts as input a data type (e.g., `np.float32` for 32-bit floating point), and returns an object containing various constants such as the largest and smallest numbers that can be represented. The code fragment below shows how to use this information, combined with (2.7) to compute the dynamic range for 32-bit floats.

```
# Get the floating point information from numpy
print('32-bit floating point\n')
float32_info = np.finfo(np.float32)
print('Smallest value:\t ', float32_info.tiny)
print('Largest value:\t ', float32_info.max)

# Compute the dynamic range by comparing max and tiny
dynamic_range = 20 * (np.log10(float32_info.max) - np.log10(float32_
↪info.tiny))
print('Dynamic range:\t {:.2f} [dB]'.format(dynamic_range))
```

```
32-bit floating point

Smallest value:          1.1754944e-38
Largest value:           3.4028235e+38
Dynamic range:           1529.23 [dB]
```

Compared to the integer representations listed above, 32-bit floats provide a substantially higher dynamic range Given the same amount of storage (32 bits), integers have a dynamic range of about 186 dB, compared to floats with 1529! This is far in excess of the dynamic range of human hearing, and often perfectly adequate for most signal processing applications. However, if for some reason 1529 decibels is not enough, we can repeat this calculation for 64-bit floats (sometimes called *double precision*).

```
# Repeat the above, but for 64-bit floats
print('64-bit floating point\n')

float64_info = np.finfo(np.float64)
print('Smallest value:\t ', float64_info.tiny)
print('Largest value:\t ', float64_info.max)

dynamic_range = 20 * (np.log10(float64_info.max) - np.log10(float64_
↪info.tiny))
print('Dynamic range:\t {:.2f} [dB]'.format(dynamic_range))
```

```
64-bit floating point

Smallest value:        2.2250738585072014e-308
Largest value:         1.7976931348623157e+308
Dynamic range:         12318.15 [dB]
```

On most modern computers and programming environments, 64-bit floating point is the default numerical representation unless you specifically request something else. All of the "continuous-valued" examples in this book use 64-bit floating point. However, many digital audio workstations (DAWs) and other audio processing software will provide the option to use 32-bit floating point because it reduces the amount of storage and computation necessary, and is still sufficient for most use cases.

2.4.6 But what does it sound like?

We can simulate quantization numerically by defining a `quantize` function as below. We'll then be able to synthesize a pure tone, and hear how it sounds when quantized to varying bit depths.

Warning: the lower bit depths can sound quite harsh.

```python
import numpy as np
from IPython.display import display, Audio

def quantize(x, n_bits):
    '''Quantize an array to a desired bit depth

    Parameters
    ----------
    x : np.ndarray, The data to quantize
    n_bits : integer > 0, The number of bits to use per sample

    Returns
    -------
    x_quantize : np.ndarray
        x reduced to the specified bit depth
    '''
    # Specify our quantization bins:
    #   2^n_bits values, evenly (linearly) spaced
    # between the min and max of the input x
    bins = np.linspace(x.min(), x.max(),
                       num=2**n_bits,
                       endpoint=False)
    return np.digitize(x, bins)
```

(continues on next page)

(continued from previous page)

```
# We'll make a 1-second example tone at 220 Hz
duration = 1
fs = 11025
freq = 220


t = np.arange(duration * fs) / fs  # Our sample times
x = np.cos(2 * np.pi * freq * t)   # The continuous signal

display('Original signal (float64)')
display(Audio(data=x, rate=fs))

# And play the audio at each bit depth
for bits in [16, 8, 4, 2, 1]:
    display('{}-bit'.format(bits))
    display(Audio(data=quantize(x, bits), rate=fs))
```

One thing you might notice from listening to these examples is that the lower quantization levels have the same fundamental frequency but sound somehow *brighter*. This is because quantization introduces discontinuities to the signal. A pure sinusoid varies smoothly and continuously, while its 1-bit quantized counterpart (a *square* wave) jumps abruptly between different sample values. Since low-frequency waves cannot change abruptly, and we can infer that severe quantization induces high frequency content in the signal. We will make this intuition more precise in a few chapters (when we cover the Fourier transform).

2.4.7   Quantization in practice

By now, we've seen how quantization works, what it looks like visually, and how it sounds. But how do we use this information practically?

16-bit quantization (65,536 distinct values) is the standard for compact disc-quality audio, and it suffices for many practical applications. 24-bit quantization is also common (16,777,216 distinct values), especially in music production or other applications with high-fidelity requirements.

Although audio data is stored this way (e.g., in .wav files), we don't usually worry about quantization when *processing* audio in a computer. Instead, the most common pattern is to convert audio signals to *floating point* numbers (which are better approximations to continuous-valued real numbers), do whatever analysis or filtering we want to do, and then only quantize again when we need to save or play back audio at the end. Usually, this happens behind the scenes, and you don't need to worry about it directly.

2.5 EXERCISES

Exercise 2.1. Imagine you are given a sampled signal $x[n]$ of N samples, taken at $f_s = 8000$ Hz. If you derive a new signal $y[n]$ by taking every other sample from $x[n]$:

$$y[n] = x[2 \cdot n],$$

- What is the sampling rate f_s of $y[n]$?

- What is the sampling period t_s of $y[n]$?

- How many samples will $y[n]$ have?

Hint. Be sure to consider what happens when N is even or odd.

Exercise 2.2. If the sampling rate is $f_s = 100$ Hz, use the aliasing equation (2.3) to find two aliases for each of the following frequencies:

- $f = 100$ Hz

- $f = -25$ Hz

- $f = 210$ Hz

Exercise 2.3. Imagine that you want to sample a continuous wave $x(t) = \cos(2\pi \cdot f \cdot t)$ of some arbitrary (known) frequency $f > 0$. Which sampling rates f_s would you need to use to produce the following discrete observations by sampling $x(t)$? Express your answers as a fraction of f.

- $x[n] = 1, 1, 1, 1, \ldots,$

- $x[n] = 1, 0, -1, 0, 1, 0, -1, 0, \ldots,$

- $x[n] = 1, \sqrt{1/2}, 0, -\sqrt{1/2}, -1, \ldots.$

Which (if any) of these sampling rates satisfy the conditions of the Nyquist-Shannon theorem?

Exercise 2.4. Using the *quantize function* from this chapter, apply different levels of quantization (varying **n_bits** from 1 to 16) to a recording of your choice. For each quantized signal, subtract it from the original signal and listen to the difference:

```
# We need to normalize the signal to cover the same range
# before and after quantization if we're going to compare them.
x_span = x.max() - x.min()
x_diff = x / x_span - quantize(x, n_bits) / 2**n_bits
```

What does it sound like for each quantization level?

## CHAPTER 3

Convolution

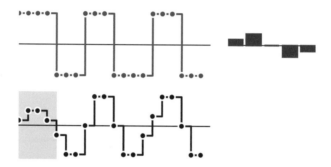

This chapter introduces the fundamental concepts behind *filtering* and **convolution**. We'll stick to the basics in this chapter and leave the more detailed analysis for later chapters.

Specifically, we'll cover:

- How to build up convolution from simpler operations of delay, gain, and mixing

- Formally defining convolution

- The concept of **impulse response**

- Different convolution **modes**

- Algebraic properties of convolution (without proof)

3.1 DELAY, GAIN, AND MIX

The term **convolution** gets thrown around quite a bit in signal processing, and it can sound more complicated than it really is. In the simplest terms, convolution consists of three basic operations:

1. **delaying** a signal by some fixed number of samples,

2. applying a **gain** to the delayed signal (changing its amplitude),

DOI: 10.1201/9781003264859-3

3. **mixing** (adding) the delayed and gained signal with the original signal.

Before we get into the equations in full generality, let's work through a couple of simple examples.

3.1.1 Example 1: delay and mix

As a first example, let's consider the case where there is no gain applied to either the delayed or original signal, so we only have to worry about delay and mixing. If our input signal is $x[n]$, and our delay is $k > 0$ samples, the output of this process will be a new signal $y[n]$ defined as

> As a general convention, we use $x[n]$ to denote an "input" signal, and $y[n]$ to denote an "output" signal, meaning the result of some computational process applied to $x[n]$.

$$y[n] = x[n] + x[n - k].$$

That is, the nth output sample $y[n]$ is the sum of the input sample $x[n]$ and the input sample k steps back in time $x[n - k]$.

Fig. 3.1 demonstrates this process for an input $x[n]$ generated by a square wave, and a delay of $k = 3$ samples.

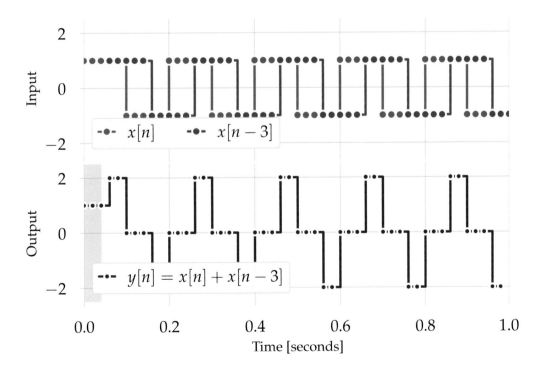

Figure 3.1 Top: a signal $x[n]$ and a delayed copy of the signal $x[n - 3]$. Bottom: the sum of the two signals.

Before we move on, there are already a couple of observations we can make.

First, the output $y[n]$ looks substantially different from the input $x[n]$. Wherever $x[n]$ and $x[n-3]$ are both positive (or both negative), $y[n]$ becomes larger (or smaller), reaching a peak amplitude of 2. When $x[n]$ and $x[n-3]$ have opposite signs, they cancel each-other out, resulting in $y[n] = 0$ (e.g., at 0.2 seconds). Overall, the resulting y signal has a different shape than the input x, more akin to a triangle wave than the square wave we started with. Different delays will produce different shapes, which we will perceive as changes in timbre.

Second, the first few samples of $y[n]$ (in the shaded region) look different from the rest of the signal: these are the only places where the value of 1 (rather than $-2, 0, 2$) occur. To understand this, we need to investigate the equation for y more carefully, and think about what happens when $n < k$. If $n < k$, then $n - k < 0$, so the sample index $n - k$ corresponds to a *negative* time index. As stated in *chapter 1*, we generally assume that a signal is silent for negative time indices (i.e., before recording started). So for the first few samples, we'll have

$$y[n] = x[n] + \cancel{x[n-k]} \quad \text{if } n < k.$$

This period of time corresponding to $n < k$ is sometimes referred to as the *warm-up* phase, where the filter we're applying has not yet seen enough of the input signal to operate completely.

Warming up and ringing out

The example in Fig. 3.1 illustrates the *warm-up* effect of convolutional filtering. Depending on how exactly the convolution operation is implemented, you may also encounter a related behavior at the end of the signal known as *ringing out*. We'll discuss this in more detail in *Convolution modes*.

Having defined the behavior for the warm-up phase, we can now translate the equation for $y[n]$ into code as follows:

```python
# Make an output buffer the same size as the input
N = len(x)
y = np.zeros(N)

# Set our delay
k = 3

for n in range(N):
    if n >= k:
        y[n] = x[n] + x[n-k]
    else:
        y[n] = x[n]
```

A complete code example is given below.

```
import numpy as np
import scipy.signal
from IPython.display import Audio

# Our input will be a 100Hz square wave for one second, sampled at␣
  ↪8 KHz.
fs = 8000
duration = 1
f0 = 100

times = np.arange(duration * fs) / fs
x = scipy.signal.square(2 * np.pi * f0 * times)

k = 8  # The delay will be 8 samples for this example
# Try changing k to see how it affects the sound of the output.
# Can you find a setting of k that makes the output silent?

N = len(x)
y = np.zeros(N)  # Initialize output buffer to match the length of x

# Compute y
for n in range(N):
    if n >= k:
        y[n] = x[n] + x[n-k]
    else:
        # At the start of the signal, x[n-k] doesn't exist yet
        y[n] = x[n]

display('Input x[n]')
display(Audio(data=x, rate=fs))

display('Output y[n] = x[n] + x[n-{}]'.format(k))
display(Audio(data=y, rate=fs))
```

3.1.2   Example 2: delay + gain

In this example, we'll mix two different delays, each with a different *gain* coefficient:

$$y[n] = \frac{1}{2}x[n] - \frac{1}{2}x[n-1].$$

Here, the delay-0 signal ($x[n]$) has a gain of $+1/2$, and the delay-1 signal ($x[n-1]$) has gain $-1/2$. Intuitively, whenever the input signal is not changing (i.e., $x[n] = x[n-1]$), then the output signal $y[n]$ should be zero. Whenever the signal is changing, the output shows the direction of the change, as seen in Figure 3.2.

Note that in this example, the gain coefficients can be either positive **or negative**.

The code below implements filter on the same square wave as the previous example. Try modifying the gain coefficients below. How does the sound change if you make both coefficients positive? Or both negative?

```python
import numpy as np
import scipy.signal
from IPython.display import Audio

fs = 8000
duration = 1
f0 = 100

times = np.arange(duration * fs) / fs
x = scipy.signal.square(2 * np.pi * f0 * times)

N = len(x)
y = np.zeros(N)

for n in range(N):
    if n >= 1:
        y[n] = 0.5 * x[n] - 0.5 * x[n-1]
    else:
        y[n] = 0.5 * x[n]

display('Input x[n]')
display(Audio(data=x, rate=fs))

display('Output y[n] = 1/2 * x[n] - 1/2 * x[n-1]')
display(Audio(data=y, rate=fs))
```

3.2 DEFINING CONVOLUTION

Now that we've worked through a couple of examples of ways to combine delay, gain, and mixing to produce effects, we can formally define the **convolution** operation. In full generality, convolution is an operation that takes two sequences x and h, and produces a new signal y. Here, x is the input signal and y is the output signal. The sequence $h[k]$ contains the coefficients by which each delayed signal is scaled before being mixed to produce the output:

- $h[0]$ scales the raw input signal $x[n]$,

- $h[1]$ scales the input delayed by one sample $x[n-1]$,

- $h[2]$ scales $x[n-2]$, and so on.

Note that x and h can have different lengths.

3.2.1 The convolution equation

Definition 3.1 (Convolution). The **convolution** of two sequences $x[n]$ (of length N) and $h[k]$ (of length K) is defined by the following equation:

$$y[n] = \sum_{k=0}^{K-1} h[k] \cdot x[n-k] \tag{3.1}$$

This equation is just a slight generalization of the examples we saw in the previous section. In words, it says that the output signal y is computed by summing together several delayed copies of the input (including delay-0), each multiplied by a coefficient $h[k]$:

$$y[n] = \sum_{k=0}^{K-1} h[k] \cdot x[n-k] = h[0] \cdot x[n-0] +$$
$$h[1] \cdot x[n-1] +$$
$$\cdots +$$
$$h[K-1] \cdot x[n-(K-1)]$$

Fig. 3.2 illustrates the computation of a convolution between a square wave and a sequence $h = [1/2, 1, 0, -1, -1/2]$. The intermediate steps show how each scaled and shifted copy of x contributes to the total convolution y (bottom subplot).

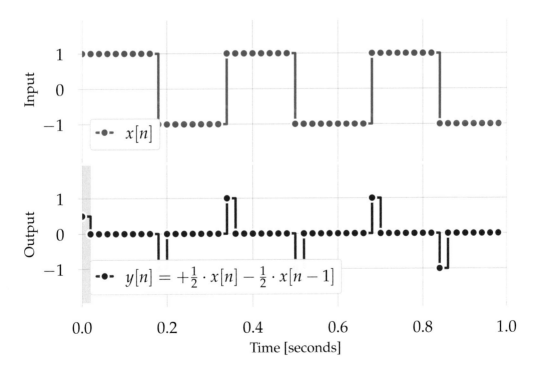

Figure 3.2 Top: an input signal $x[n]$. Bottom: the output signal.

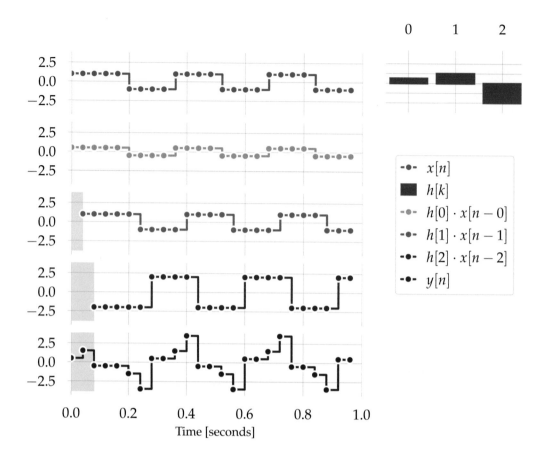

Figure 3.3 Top-left: a signal x is delayed and scaled by each coefficient of the filter $h = [1/2, 1, -2]$ (top-right). Left, rows 2–4: each delayed and scaled copy of x is added together to form the output signal $y[n]$ (bottom-left).

This operation is so common in signal processing that it gets its own notation:

$$y = h * x, \tag{3.2}$$

which is short-hand for the summation above, and we read as y *is x* **convolved** *with* h.

Incidentally, this is why we use $a \cdot b$ to denote scalar (number) multiplication instead of $a * b$. As we will see in *chapter 10*, convolution can be viewed abstractly as a kind of "multiplication between signals", and using the same $*$ notation for both operations could lead to confusion.

3.2.2 Implementing convolution

Convolution can be implemented in a variety of ways. They all produce the same result, but some implementations may be more or less efficient, or easy to understand. The first implementation we'll cover is the one used to create Fig. 3.3: we compute the contribution due to each delay individually, and accumulate them in the output buffer y:

```
# Compute the lengths of our input and filter
N = len(x)
K = len(h)

# Allocate an output buffer
y = np.zeros(N)

# Iterate over delay values
for k in range(K):

    # For each delay value, iterate over all sample indices
    for n in range(N):

        if n >= k:
            # No contribution from samples n < k
            y[n] += h[k] * x[n-k]
```

One thing to observe is that y[n] is the sum of many pairs of numbers multiplied together. Because addition does not depend on the order in which arguments appear (e.g., $3 + 4 = 4 + 3$), we can compute this sum in any order we like. In particular, we can change the order of the loops: the outer loop can range over samples, and the inner loop can range over delay values:

```
for n in range(N):
    for k in range(K):
        if n >= k:
            y[n] += h[k] * x[n-k]
```

This implementation computes exactly the same y[n] as the first one, but it does so **one sample at a time**. This approach is more common in practice because it allows you to perform convolution in real time, as signals are being generated! The first example would not allow this, because it has to scan the entire input signal x for one delay value before moving on to the next.

There are yet more ways to implement convolution, but we'll have to wait until *chapter 10* to see how those work. Once you're comfortable with the idea of convolution, it's recommended to not implement it yourself, but instead to use a pre-existing implementation like np.convolve.

3.2.3 Special cases

Before going deeper into the theory of convolution, let's step back and see how the idea encompasses familiar operations as special cases.

Delaying by d samples

In a signal delay, there is no gain, and effectively no *mixing* either. All we want to do is displace the signal in time by a fixed number d of samples (Figure 3.4), producing

$$y[n] = x[n - d].$$

This is done by setting h to be an array of d zeros followed by a single 1:

$$h = [\underbrace{0, 0, \ldots, 0}_{d \text{ zeros}}, 1]$$

This h has $K = d+1$ coefficients. Plugging this h into the convolution equation, we'd see that all terms are 0 except for the last (corresponding to $h[K - 1] = 1$), which results in

$$
\begin{aligned}
y = h * x \quad \Rightarrow \quad y[n] &= h[K - 1] \cdot x[n - (K - 1)] \\
&= h[d + 1 - 1] \cdot x[n - (d + 1 - 1)] \\
&= h[d] \cdot x[n - d] \\
&= x[n - d],
\end{aligned}
$$

which is exactly what we want: $y[n] = x[n - d]$.

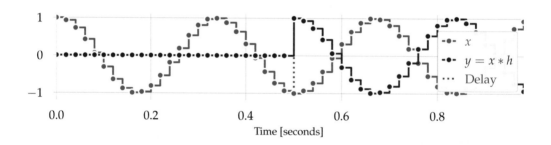

Figure 3.4 Delay can be implemented by convolution.

Gain

Applying gain to a signal means that all samples $x[n]$ is scaled by some fixed amount G, as seen in Figure 3.5, so

$$y[n] = G \cdot x[n].$$

This can be implemented by setting h to be an array with only a single element (corresponding to a delay of 0):

$$h = [G].$$

Again, plugging into the convolution equation, the summation has only one term:

$$y = h * x \quad \Rightarrow \quad y[n] = h[0] \cdot x[n - 0]$$
$$= G \cdot x[n].$$

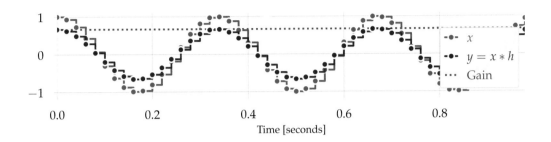

Figure 3.5 Gain can be implemented by convolution. This example uses a gain value of 2/3 to reduce the amplitude of the input signal x.

Using convolution to implement gain (or delay) might be overkill, but it's helpful to understand these cases to build an intuition for the properties of convolutional filters.

3.3 IMPULSE RESPONSE

The convolution operation is closely related to the idea of an **impulse response**. In this section, we'll work through what this all means, and how convolution can be related to acoustic wave propagation.

3.3.1 What is an impulse?

Before we go further, we'll need to define an **impulse**. An impulse is an idealized signal consisting of a single 1, followed by (in theory at least) infinitely many zeros:

$$x_| = [1, 0, 0, 0, \cdots].$$

Impulses are theoretical constructs, and cannot exist in nature. The closest familiar sounds to an impulse would be things like a balloon popping, or tapping two hard objects together, but these sounds will only approximate an ideal impulse.

We can construct impulses inside a computer (or with a pencil and paper), and doing so can help us understand the behavior of many signal-processing operations.

3.3.2 Impulse response of a filter

In general, a *filter* is any process that consumes one signal as input and produces a new signal as output. One might express this notationally as

$$y = g(x)$$

for input signal x and filter operation g.

The **impulse response** of a filter g is the signal y produced by applying g to an impulse:

$$g\left(x_|\right).$$

This is a broad and abstract definition, but in casual conversation, when people refer to filters they most often mean *linear filters*. We'll go one step further in this section, and assume that the filtering operation is a convolution between the input x and some fixed sequence h. One may then ask *what is the impulse response of a convolutional filter?*

Here, we'll go back to the definition of convolution:

$$y[n] = \sum_{k=0}^{K-1} h[k] \cdot x[n-k].$$

Plugging in our definition of an ideal impulse $x = x_|$, we see that $x[n-k] = 1$ if $n = k$ (so that $n - k = 0$) and $x[n-k] = 0$ otherwise. This means that we can simplify the calculation significantly:

$$\begin{aligned}
y[n] &= \sum_{k=0}^{K-1} h[k] \cdot x[n-k] \\
&= h[n] \cdot x_|[0] \\
&= h[n].
\end{aligned} \tag{3.3}$$

That is, the impulse response of a convolutional filter is the sequence h itself.

Put another way, the impulse response alone is enough to completely characterize a convolutional filter: no other information is necessary.

3.3.3 Finite impulse response (FIR)

You may have heard the term **finite impulse response (FIR)** and wondered what it meant. In plain terms, an FIR filter is any filter whose impulse response goes to 0 and stays there after some finite number of samples. In general, this critical number of samples is a property of the filter in question, and will vary from one filter to the next.

In the case of convolutional filters, $g(x) = h * x$ (for some finite sequence h of length K), the impulse response must go to 0 after K samples. This is because any output sample after that point will depend only on the trailing zeros in the impulse.

In short, **convolutional filters have a finite impulse response.**

In later chapters, we'll see examples of other kinds of filters which use feedback to achieve an *infinite impulse response (IIR)*. But for now, there's still much more to explore with convolutional filters.

3.3.4 Room impulse response

Beyond digital filters, you can also think about impulse responses of physical environments. Imagine placing a sound source and a microphone in a room, and for now, let's assume that the room's walls have perfectly (acoustically) reflective surfaces. Any sound emanating from the source will then have multiple paths to the microphone: the direct (shortest) path, as well as longer paths that reflect from each wall (or multiple walls). If the sound source produces an ideal impulse, we can observe the impulse arriving at the microphone at different *delay* times (corresponding to the different paths of arrival). Because each path has different length, the intensity of the sound will diminish for the longer paths corresponding to higher delay times.

This process is illustrated in Figure 3.6, for an example where we have two perpendicular walls (and no floor or ceiling, just to keep things simple).

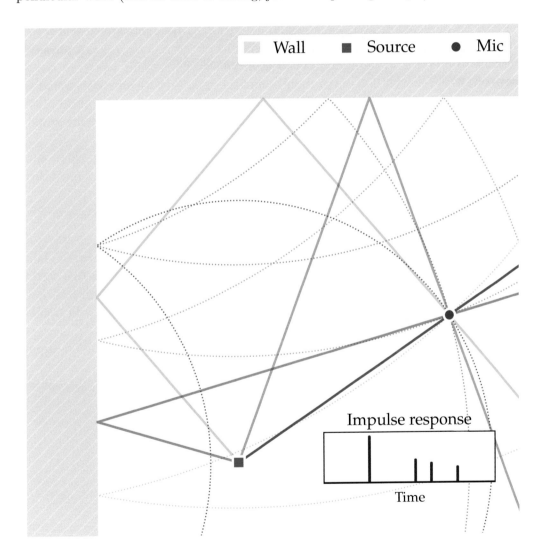

Figure 3.6 A sound radiating from a source (square) emanating in all directions reflects off surfaces, resulting in multiple paths (straight lines) to a recording device. Dashed circles indicate the wavefronts at different times, and darker lines indicate higher intensity.

This entire physical process can be thought of as implementing a convolution. Each reflection path corresponds to a different delay k, and the decrease in recorded intensity corresponds to the gain coefficient $h[k]$ for that path.

Note that the impulse response depends not only on the physical environment, but also the positioning of the sound source and microphone: if you move either (or both) of these, the impulse response will generally change.

The example above is significantly simplified from physical reality in several ways. First, a real room would have three dimensions (floors and ceilings) which can also provide reflective surfaces and increase the number of paths. Second, the surface materials play a large part in how sound is reflected or diffused when colliding with a wall, so the observed delayed signal would not generally be a perfectly scaled copy of the impulse. Third, as mentioned at the beginning of this section, it's not physically possible to produce an *ideal* impulse, so what you would actually record at the microphone is the convolution of the room's impulse response with whatever sound was actually produced by the source. (In practice, non-ideal impulses can also be used, as can other signals such as sinusoidal sweeps, but that's a bit beyond the scope of this section.)

Despite the limitations of this example, it can still be instructive to think about convolution as a physical process, as it provides mechanisms which implement delay, gain, and mixing in a natural context.

What can I do with this?

Say you were able to perform the above experiment in an environment of your choice, resulting in an impulse response recording h. Remember how earlier in this section, we saw that the impulse response of a convolution $h * x$ is just h? This means that if we have h, we can now apply it to **any signal** x (not just an impulse), and simulate the effect of hearing it in the environment characterized by h.

The following code example demonstrates this process for a pre-recorded impulse response and input signal.

```python
import numpy as np
from IPython.display import display, Audio

# We'll need soundfile to load waves
import soundfile

# Input signal is a short piano excerpt
# https://freesound.org/people/Piotr123/sounds/511749/
x, fs = soundfile.read('511749__piotr123__jazz-piano-intro-mono.wav')

# Impulse response is from a church:
# https://freesound.org/s/474296/
h, fs2 = soundfile.read('474296__petherjwag__ir-church-01-mono.wav')
```

(continues on next page)

(continued from previous page)

```
# Check that the sampling rates match
assert fs == fs2

# Convolve the signal with the impulse response
y = np.convolve(x, h)

display('Impulse response')
display(Audio(data=h, rate=fs))
display('Input signal')
display(Audio(data=x, rate=fs))
display('Output signal')
display(Audio(data=y, rate=fs))
```

I encourage you to try this out for yourself. If you don't have the materials necessary to record your own impulse responses, there are plenty of examples available on Freesound.org.

3.4   CONVOLUTION MODES

As we saw in the first section of this chapter, the first few samples produced by a convolution $y = x * h$ depend on *negative sample indices* $x[n - k]$ where $n < k$. To ensure that the convolution is well-defined for these indices, we assume $x[n - k] = 0$ for $n < k < K$ (the length of the filter), which we may also think of as **padding** the signal with zeros on the left. This means that we can think of the first k samples of y as depending on these additional zeros. Any sample $n \geq k$ will not depend on this padding.

Depending on how you want to use convolution, it may or may not be appropriate to include these first few samples. In this section, we'll describe different commonly used **modes** of convolution, which can all be implemented by some combination of padding and trimming the inputs and outputs. Note that these modes do not fundamentally alter how convolution works: all that changes is which outputs are retained in the output signal.

Throughout this section, we will assume that x and h have N and K samples, respectively, and that $N \geq K$. This latter assumption is for convenience, and as we'll see in the next section, it is not necessary.

3.4.1   Full mode

Full mode convolution is implemented by padding the input on both the left *and the right* by $K - 1$ samples each. A full mode convolution will have $N + K - 1$ output samples.

The intuition for this is that it captures both a *warm up* phase ($n < K$), as well as a *ringing out* phase ($N \leq n < N + K$). The warm up phase contains all outputs which depend on some (but not only) prepended silence, the ringing out phase similarly contains outputs which depend on some (but not only) appended silence.

> In the context of audio effects, the warm up phase can be related to *latency*: how much time displacement is induced by the convolution? Similarly, the ringing out phase can be related to *reverb tails*: how long do we have to let the signal play out before it settles back down to silence?

These effects can be seen in the example below, where we compute a full convolution between an increasing sequence $[5, 6, 7, 8, 9]$ and a differencing filter $[1, 0, -1]$. With this example, we'll have $y[n] = x[n] - x[n-2]$ in general, which will be the constant 2 for this sequence. However, at the beginning and end, you can see the effects of left- and right-padding by zeros.

```
# Full mode example
x = np.arange(5, 10)

# This is a differencing filter:
#    y[n] = x[n] - x[n-2]
h = np.array([1, 0, -1])

y = np.convolve(x, h, mode='full')

print('Input            :\t', x)
print('Filter           :\t', h)
print('Full convolution:\t', y)
```

```
Input           :        [5 6 7 8 9]
Filter          :        [ 1   0 -1]
Full convolution:        [ 5   6   2   2   2 -8 -9]
```

3.4.2 Valid mode

Valid mode convolution eliminates any outputs that depend on padding at all. If we start from the **full mode** convolution described above, the *valid* outputs are the middle 3, because they depend only on values explicitly observed in the input: $7 - 5$, $8 - 6$, and $9 - 7$.

A valid mode convolution will have $N - K + 1$ output samples.

```
y = np.convolve(x, h, mode='valid')
print('Input            :\t', x)
print('Filter           :\t', h)
print('Valid convolution:\t', y)
```

```
Input           :           [5 6 7 8 9]
Filter          :           [ 1  0 -1]
Valid convolution:          [2 2 2]
```

3.4.3 Same mode

Same mode is derived from **full mode** by extracting N samples from the full convolution, so that the output y has the *same length* N as x. Most implementations of **same mode** extract the *middle* of the output, so that both the first and last $K/2$ samples depend on zero padding, as demonstrated below:

```python
y = np.convolve(x, h, mode='same')

print('Input           :\t', x)
print('Filter          :\t', h)
print('Same convolution:\t', y)
```

```
Input           :           [5 6 7 8 9]
Filter          :           [ 1  0 -1]
Same convolution:           [ 6  2  2  2 -8]
```

Sometimes, you may want a left-aligned (or right-aligned) same-mode convolution, including the warm-up but not ring-out (or vice versa). These can be helpful if you want to introduce a delay effect to a signal, but retain its length. While the standard implementation used above does not do this, it's easy enough to construct by trimming the full mode convolution differently:

```python
# Include only the warm-up phase
N = len(x)
y = np.convolve(x, h, mode='full')
y = y[:N]  # Truncate to the first N samples

print('Input           :\t', x)
print('Filter          :\t', h)
print('Same (left-align):\t', y)
```

```
Input           :           [5 6 7 8 9]
Filter          :           [ 1  0 -1]
Same (left-align):          [5 6 2 2 2]
```

```python
# Discard the warm-up, but include the ring-out
N = len(x)
y = np.convolve(x, h, mode='full')
y = y[-N:]  # Truncate to the last N samples
```

(continues on next page)

(continued from previous page)

```
print('Input              :\t', x)
print('Filter             :\t', h)
print('Same (right-align):\t', y)
```

```
Input             :        [5 6 7 8 9]
Filter            :        [ 1  0 -1]
Same (right-align):        [ 2  2  2 -8 -9]
```

3.4.4 Summary

The different convolution modes are summarized visually in Figure 3.7, with h as the differencing filter used above, and x now as a square wave over 25 samples.

The shaded regions of the bottom subplot correspond to those which depend on zero-padding.

The table below provides a rubric for deriving the different convolution modes from **full** mode in Python. For each mode, the length of the output signal is listed, as well as the *slice notation* for finding the different mode outputs within the `full` convolution.

In this table, we've assumed that $N \geq K \geq 1$.

Mode	Length	Slice	Code
Full	$N + K - 1$	`0 : N + K - 1`	`np.convolve(x, h, mode='full')`
Valid	$N - K + 1$	`K : N`	`np.convolve(x, h, mode='valid')`
Same (left)	N	`0 : N`	`np.convolve(x, h, mode='full')[:N]`
Same (centered)	N	`(K-1)//2 : N + (K-1)//2`	`np.convolve(x, h, mode='same')`
Same (right)	N	`K -1 : N + K-1`	`np.convolve(x, h, mode='full')[K-1:]`

Remember that array slices in Python do not include the upper limit. For example, `x[K:N]` denotes indices `x[K]`, `x[K+1]`, `x[K+2]`, … `x[N-1]`.

3.5   PROPERTIES OF CONVOLUTION

As stated earlier in this chapter, convolution acts like a kind of abstract *multiplication* between signals. Making this statement more precise will have to wait until we've developed the Fourier transform, but for now we can end this chapter by showing some of the properties that convolution shares with multiplication.

Note: we'll assume for now that all convolutions are **full mode**. With appropriate trimming, you can extend these properties to the other modes, but the notation will get a bit more complicated.

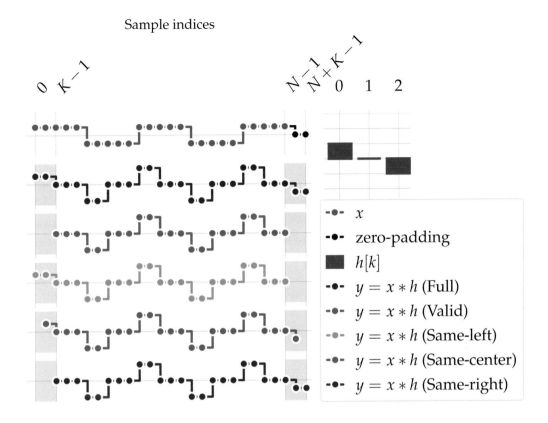

Figure 3.7 Top-left: an input signal $x[n]$, followed by zero-padding. Top-right: the coefficients of a filter $h[k]$. Remaining left plots: The full convolution $y = h * x$ (dots) can be trimmed in various ways to produce each of the different convolution modes. The shaded regions correspond to output sample positions in the warm-up and ringing-out phases.

3.5.1 Commutative property

Convolution is **commutative**, which is a fancy word for saying that order doesn't matter. Precisely, commutativity is the property that for any x and h, the following holds:

Property 3.1 (Commutativity of convolution). If x is a signal and h is an impulse response, then

$$h * x = x * h.$$

What this says in words is that there is no difference between *signals* and *filters*: any sequence of numbers can be interpreted as either (or both)!

3.5.2 Associative property

Convolution is also **associative**, which means that if we're chaining a series of convolutions together, it doesn't matter which one you do first. Imagine that we convolve

$x * h$ and then convolve the result by some other filter g. Associativity says that we could equivalently convolve h with g first, and then convolve the result with x. Formally, we have

Property 3.2 (Associativity of convolution). If x is a signal and h and g are impulse responses, then

$$g * (h * x) = (g * h) * x$$

Why does this matter? Imagine that h and g are both short (maybe a dozen samples) and x is long (millions of samples). Computing $h * x$ would take work proportional to the length of x, as would convolving the output with g: you'd have to step through large arrays twice. However, if you convolve $g * h$ first, that only takes work proportional to the length of the filters (which are small), and then you'd only have to step through the large array once. Reasoning about filters in this way can help you develop more efficient processing chains without changing the calculations.

By combining associativity with commutativity, it turns out that a sequence of convolutions $x * h * g * \ldots$ can be equivalently computed **in any order!** This fact can be exploited to find the most efficient possible order (e.g. by ordering the inputs from short to long).

3.5.3 Distributive over addition

Imagine we have two signals x_1 and x_2 that we'd like to filter by h and then mix. The **distributive** property says that we can equivalently mix the signals first, and then convolve by h:

Property 3.3 (Distributivity of convolution). If x_1, x_2 are signals and h is an impulse response, then

$$h * x_1 + h * x_2 = h * (x_1 + x_2).$$

Like the associative property, the distributive property can be used to reduce the amount of computation needed to produce a particular output. Note that the left-hand side of the equation has two convolutions, and the right-hand side has only one.

As we'll see later, this property can be generalized slightly to show that convolution is a *linear operator*.

3.5.4 Proving these properties

To prove that one of these properties holds, one would generally proceed by using the formal definition of convolution, and then applying a sequence of algebraic manipulations until we arrive at the desired result.

As an example, here's how one could prove the distributive property (*Property 3.3*).

Proof. Let $y = h * (x_1 + x_2)$. Then the nth sample of y is

$$y[n] = \sum_{k=0}^{K-1} h[k] \cdot (x_1[n-k] + x_2[n-k]) \qquad \text{by definition of } y$$

$$= \sum_{k=0}^{K-1} (h[k] \cdot x_1[n-k] + h[k] \cdot x_2[n-k]) \qquad \text{pull } h[k] \text{ into sum}$$

$$= \left(\sum_{k=0}^{K-1} h[k] \cdot x_1[n-k] \right) + \left(\sum_{k=0}^{K-1} h[k] \cdot x_2[n-k] \right) \qquad \text{re-arrange the sum}$$

$$= (h * x_1)[n] + (h * x_2)[n] \qquad \text{by definition of } * .$$

And since this holds for all n, we get

$$h * (x_1 + x_2) = y = (h * x_1) + (h * x_2).$$

□

Similar types of arguments to the one above can be used to show that commutativity or associativity hold. For the mathematically inclined reader, this can be a good exercise. However, **this style of proof can be tedious.** For the time being, we'll assume that these properties hold without giving explicit proofs.

In *chapter 10*, we'll see some mathematical tools that make it much simpler to show these properties of convolution from (more or less) first principles, without relying on so much algebra.

3.6 LINEARITY AND SHIFT-INVARIANCE

We'll end this chapter with a discussion of two of the more advanced properties of convolutional filters: linearity and shift-invariance. These are general concepts which can be applied to more than just convolutional filters, but for now, we'll focus on the specific case of convolution.

Taken together, these two conditions (linearity and shift-invariance) are often referred to as *LSI*, and a system satisfying these properties is called an *LSI system*.

These properties might seem abstract for now, but they will become important later on when we need to analyze the behavior of filters by using the Fourier transform.

3.6.1 Shift-invariance

The idea of shift-invariance is that delay can be applied either before or after a system g, and produce the same result.

Formally, let Δ denote a d-step delay filter for some fixed but arbitrary delay $d \in \mathbb{N}$:

$$\Delta = [\underbrace{0, 0, \ldots, 0}_{d \text{ times}}, 1]$$

so that convolving a signal x with Δ yields

$$(\Delta * x)[n] = x[n - d]. \tag{3.4}$$

Definition 3.2 (Shift invariance). A filter g is **shift-invariant** if for all delays d (implemented by delay filter Δ) and all input signals x, the following is true:

$$g(\Delta * x) = \Delta * g(x).$$

In plain language, equation *Definition 3.2* says that if we delay x by d samples, and then process the delayed signal by the system g, we will get the same result as if we had first applied g to x (without delay) and then delayed the result.

> **Shift-invariance or time-invariance?**
>
> The term *time-invariant* is also commonly used to describe the property above. There are two reasons why shift-invariance is a more precise term:
>
> 1. "Time-invariant" does not describe what transformation of time the system is invariant to. Shifting is one transformation, but so is stretching (e.g., changing units from seconds to milliseconds), reversal, or any number of other actions.
>
> 2. "Shift-invariant" can be applied to systems operating on other kinds of signals, such as images, which may have different notions of "position" than time (e.g., pixel coordinates).
>
> For these reasons, we'll stick to the more precise terminology, but you are likely to encounter the *time-invariant* terminology elsewhere, so remember that these are the same thing.

Theorem 3.1 (Convolution is shift-invariant). Let $g(x) = h * x$ be a convolutional system for some impulse response h and any input signal x.

Any such $g(x)$ is shift-invariant, and satisfies *Definition 3.2*.

Proof. We need to verify that

$$g(\Delta * x) = \Delta * g(x).$$

Using the assumption that g is convolutional, we can appeal to the associative and commutative properties stated in the previous section:

$$
\begin{aligned}
g(\Delta * x) &= h * (\Delta * x) && \text{apply } g \text{ to delayed input} \\
&= (h * \Delta) * x && \text{associative rule} \\
&= (\Delta * h) * x && \text{commutative rule} \\
&= \Delta * (h * x) && \text{associative rule} \\
&= \Delta * g(x) && \text{definition of } g.
\end{aligned}
$$

$\square$

This gives us a shortcut to showing shift-invariance: if you can implement a system g in terms of convolution, it automatically satisfies shift-invariance.

3.6.2 Linearity

Linearity is another important characteristic of many systems, including convolution. Broadly speaking, it encapsulates our notions of *gain* and *mixing* of signals.

Definition 3.3 (Linearity). A system g is **linear** if for any two signals x_1 and x_2, and for any pair of numbers c_1 and c_2, the following holds:

$$g(c_1 \cdot x_1 + c_2 \cdot x_2) = c_1 \cdot g(x_1) + c_2 \cdot g(x_2).$$

This might look like a tangle of symbols, so it helps to think about a few special cases. For instance, if we take an example where $x_2[n] = 0$ is a silent (all zeros) signal, and $c_2 = 0$, then we get the simpler case

$$g(c_1 \cdot x_1) = c_1 \cdot g(x_1),$$

which is also known as *homogeneity*. This says that gain can be applied before or after the system g with no difference in the output.

Similarly, if we take $c_1 = c_2 = 1$, then we get

$$g(x_1 + x_2) = g(x_1) + g(x_2),$$

which is also known as *additivity*. This says that signals can be mixed before applying g, or after, and the result will be the same. We've already seen this particular case before, when discussing how convolution *distributes over addition*.

Definition 3.3 combines these two ideas into one compact form.

Aside: why *linear*?

In the context of signal processing, the term *linearity* might seem a little out of place, if not downright mysterious.

Like many parts of our mathematical vocabulary, the term is borrowed from the theory of *linear algebra*, where operators like rotation or translation are defined to ensure that the shapes of straight lines are preserved after application.

The idea is much more general though, and in our context, *linearity* should be interpreted as preserving *gain* and *mixture* of signals.

Theorem 3.2 (Convolution is linear). Let $g(x) = h * x$ be a convolutional system for some impulse response h.

Then $g(x)$ satisfies *Definition 3.3* and is a linear system.

To prove *Theorem 3.2*, let h be an impulse response, and let x_1 and x_2 denote arbitrary signals, and c_1, c_2 denote arbitrary numbers.

We could prove that g is linear by brute force, using the definition of convolution and working through the algebra directly. Many people will find this exercise somewhat tedious, but it is possible.

We'll take a slightly different route here though, relying on facts we already know about convolution.

Proof. Recall that we can implement gain as a convolutional system by creating an impulse response with only one element: *Gain*. Let $C_1 = [c_1]$ and $C_2 = [c_2]$, so that

$$c_1 \cdot x_1[n] = (C_1 * x_1)[n]$$
$$c_2 \cdot x_2[n] = (C_2 * x_2)[n].$$

Then, we can show linearity in terms of associative, commutative, and distributive properties:

$$
\begin{aligned}
g(c_1 \cdot x_1 + c_2 \cdot x_2) &= h * (c_1 \cdot x_1 + c_2 \cdot x_2) && \text{apply } g(x) = h * x \\
&= h * (C_1 * x_1 + C_2 * x_2) && \text{use definition of } C_1, C_2 \\
&= (h * C_1 * x_1) + (h * C_2 * x_2) && \text{distributive rule} \\
&= (C_1 * h * x_1) + (C_2 * h * x_2) && \text{commutative rule (twice)} \\
&= C_1 * (h * x_1) + C_2 * (h * x_2) && \text{associative rule (twice)} \\
&= C_1 * g(x_1) + C_2 * g(x_2) && \text{definition of } g \\
&= c_1 \cdot g(x_1) + c_2 \cdot g(x_2) && \text{definition of } C_1, C_2
\end{aligned}
$$

□

Note that the use of convolution to implement gain here is mainly for convenience: having all operations expressed as either sums or convolutions buys us a bit of notational simplicity. That said, it is still completely valid to prove this from first principles, but the derivation would be longer.

3.6.3   Systems which are not LSI

Not all filtering operations you might want to do satisfy the LSI conditions. Some might be only linear, some might be only shift-invariant, and some might be neither. Identifying these properties can be helpful for understanding both how a particular system behaves, and how it can be combined with others. In particular, since all convolutional systems are LSI, we can infer that any system which is *not* LSI cannot be implemented by a convolution.

There are many systems which are shift-invariant but not linear. Generally speaking, any system which operates on each sample *value* independently (like *gain*) will be shift-invariant. Linearity will then depend on *how* the sample values are processed.

Clipping

Clipping is a key step of many *distortion* effects, leading to the distinctive sound of distorted electric guitar. This is because clipping models what happens to amplifier circuits when pushed past their limits (overdriven): the amplifier is asked to produce a voltage higher than it's capable of, and the output signal saturates at its voltage limits.

Example 3.1 (Clipping). A **clipping** system limits the output of a system so that it cannot be less than some minimum value v_- or greater than a maximum value v_+, (see Figure 3.8). In equations, this looks like:

$$y[n] = \begin{cases} v_+ & x[n] \geq v_+ \\ v_- & x[n] \leq v_- \\ x[n] & \text{otherwise} \end{cases}.$$

Equivalently, in code this is expressed as:

```python
def clip(x, vmin, vmax):
    # This behavior is implemented by np.clip,
    # but we provide a full implementation
    # here for reference.

    N = len(x)
    y = np.zeros(N)

    for n in range(N):
        if x[n] >= vmax:
            y[n] = vmax
        elif x[n] <= vmin:
            y[n] = vmin
        else:
            y[n] = x[n]
    return y
```

Remember that if a system is linear, it must be linear for *all* signals x_1, x_2 and scalars c_1, c_2. To show that clipping is not linear, we only need to find one counter-example case: a setting of x_1, x_2, c_1, c_2 for which it fails to hold.

Imagine taking a signal $x_1[n] = v_+ > 0$ (for all n), letting $x_2 = 0$, and setting $c_1 = 2$. Then

$$g(c_1 \cdot x_1) = g(2 \cdot v_+) = v_+$$

but this is not equal to $c_1 \cdot g(x_1) = 2 \cdot v_+$. We can then conclude that the system cannot be linear because it does not preserve gain.

In general, showing that a system is not linear requires some insight about how the system operates. In this case, we exploited the fact that the behavior of g changes when the input is above v_+, and used that to construct a counter-example. This type of argument often seems obvious in hindsight, but creating counter-examples is a skill that takes practice to develop.

Just as in the previous case, there are many systems which are *linear*, but not *shift-invariant*. In this case, the dependence to look at is on the sample indices.

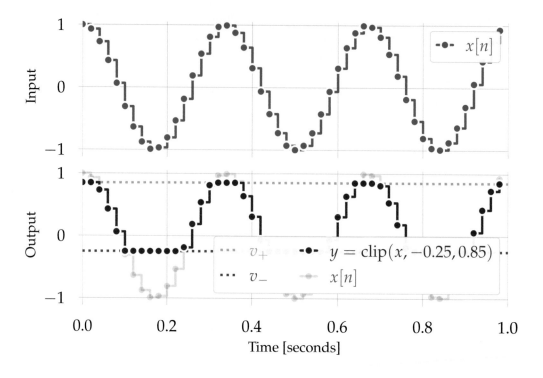

Figure 3.8 **Top**: an input signal x. **Bottom**: the result of clipping x to lie between v_- and v_+ (dashed lines).

Example 3.2 (Time-reversal). A *time-reversal* system does exactly what it sounds like: it plays the signal backward. Mathematically, this is done by swapping the sample at index n with the one at index $N - 1 - n$:

$$y[n] = x[N - 1 - n]$$

or in code,

```
def reverse(x):
    # This could be equivalently done with the one-liner:
    #    return x[::-1]

    N = len(x)
    y = np.zeros(N)

    for n in range(N):
        y[n] = x[N-1-n]
    return y
```

Note that the previous example (clipping) can operate independently on each sample, and it does not need access to the entire signal at once to operate correctly. However, the time-reversal system does need to see the entire signal in advance.

To show that time-reversal is not shift-invariant, we again must construct a counter-example. Here the key property that we're looking at is symmetry (in time), so we should probably consider signals which do not look the same forwards as backwards. Sinusoids and square waves probably won't work, but a sawtooth will do nicely.

After that, we need only check what happens when we delay the input by some number of samples. Here, we'll take a 5-step delay, but any positive number would do.

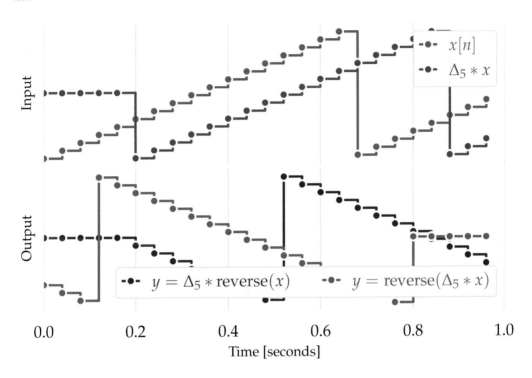

Figure 3.9 **Top**: a signal $x[n]$ and a 5-sample delayed copy. **Bottom**: applying time-reversal before or after delaying the input produces different output signals.

Fig. 3.9 shows that changing the order of operations (delay first or reverse first) matters, since the two signals in the bottom plot are not identical. From this example, we can conclude that time-reversal is *not* shift-invariant.

3.6.4 Summary

What we've seen here is that all convolutional systems are both linear and shift-invariant, and if a system fails to satisfy either of these conditions, then it cannot be implemented by convolution.

The connection here is even stronger than that: *all LSI systems are convolutions as well.* Proving that this is true is beyond the scope of this text and requires some slightly more sophisticated mathematics. However, the equivalence between convolutions and LSI systems is one of the most powerful concepts in all of signal processing.

3.7 EXERCISES

Exercise 3.1. For each of the following systems, what is its impulse response?

1. $y[n] = x[n] - x[n-1]$

2. $y[n] = 2 \cdot x[n] + x[n-2]$

3. $y[n] = \frac{1}{3} \cdot (x[n-2] + x[n-3] + x[n-4])$

Exercise 3.2. If x is a signal of $N = 100$ samples, an h is an impulse response of $K = 11$ samples, how many samples does $y = h * x$ have in each of the following modes?

1. Full

2. Valid

3. Same (centered)

Exercise 3.3. For each of the following systems, determine whether it is linear, shift-invariant, both, or neither.

1. $y[n] = -x[n]$

2. $y[n] = x[0]$

3. $y[n] = \frac{1}{2} \cdot |x[n] + x[n-1]|$

4. $y[n] = 20$

5. $y[n] = n^2$

Exercise 3.4. Let $h = [1/K, 1/K, \ldots, 1/K]$ (K times) denote a *moving average filter* so that if $y = h * x$, then $y[n]$ is the average of the previous K samples in x.

Using an input recording x of your choice (not more than 10 seconds), compute y for different values of $K = 1, 4, 16, 64, 256, 1024$ and listen to each one.

How does each y sound compared to the original signal? Do you notice any artifacts?

```
# Starter code for making a moving average filter
K = 8
h_K = 1/K * np.ones(K)
```

## CHAPTER 4

Complex numbers

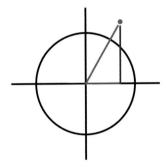

This section covers the fundamentals of complex numbers, and should help give a sense of how and why they're useful for modeling periodic phenomena such as audio signals.

Why use complex numbers?

Complex numbers were originally developed to solve polynomial equations (e.g., find x such that $x^2 + x + 1 = 0$), and it's true that many of the things we do in signal processing involve exactly these kinds of equations. However, this isn't a terribly satisfying motivation for using complex numbers to study signal processing, and this is the point at which many students start to get confused about why we need all of this math.

There is, however, a much deeper reason why complex numbers arise in signal processing, and it has to do with the link between arithmetic and geometry.

Think about what happens when you multiply one real number x by another y (Figure 4.1). If $y > 1$, then $x \cdot y$ becomes bigger than x, which you can think of as *stretching* the number line. If $y < 1$, then $x \cdot y$ becomes smaller, which embodies the opposite effect, *compressing* the number line. And if $y < 0$, then the sign of $x \cdot y$ flips from x's (positive becomes negative, and vice versa), which you can think of as *mirroring* the number line horizontally. Stretching, compressing, and reflecting are all familiar **geometric** operations that happen to correspond to the arithmetic of real numbers.

But what if the numbers are complex instead of real? It turns out that you still get the same basic types of operations (stretching, compressing, and reflecting),

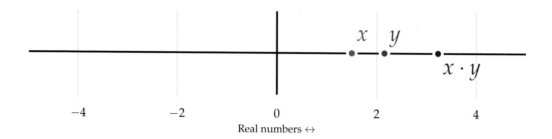

Figure 4.1 Multiplication on the real number line acts like stretching, compressing, or reflecting across the 0 point.

but now in *two* dimensions instead of 1. But there's an additional **twist**: complex multiplication also **rotates** points in the plane. Rotation is a one of the simplest models of periodicity, just ask any clock-maker.

Tip. Rotation models **repetition**.
Complex multiplication models **rotation**.

4.1 DEFINING COMPLEX NUMBERS

Recall that there is no **real** number x such that $x^2 = -1$. However, if we *imagine* a "number" j (which need not be **real**) such that $j^2 = -1$, it turns out that this leads to a whole new type of numbers, which we call **complex**.

As a general convention, to avoid confusion, it is common to denote real numbers by x and y and complex numbers by z and w.

Formally, a complex number z consists of a real part and an imaginary part. We can equivalently think of the imaginary part as being a real number multiplied by j, so that the complex number can be expressed succinctly as

$$z = a + jb.$$

You can think of a as "how much real is in z", and b as "how much imaginary is in z". If $a = 0$, then $z = jb$ is called **purely imaginary**. Likewise, if $b = 0$, then $z = a$ is called **purely real**.

In Python, we can represent complex numbers using a similar notation:

Tip. It does not work to say
```
z = 1 + j
```
because Python will interpret j as a variable if there isn't a number in front of it.

Instead, say z = 1 + 1j.

```
z = 1 + 2j
print(z)
```

```
(1+2j)
```

We can then extract their real and imaginary parts by accessing `z.real` and `z.imag`:

```
print('Real part:        {}'.format(z.real))
print('Imaginary part:  {}'.format(z.imag))
```

```
Real part:        1.0
Imaginary part: 2.0
```

In mathematical notation, you may also see Rez and Imz used to denote the real and imaginary parts of a complex number z.

This representation of z is known as the **rectangular form**, as you can interpret the real and imaginary parts as length and width of a rectangle. Interpreting complex numbers in this way leads to a natural way to reason about them as points in a two-dimensional plane, where the horizontal position measures the real part, and the vertical axis measures the imaginary part. Fig. 4.2 demonstrates this idea, and shows the position of a few selected complex numbers z.

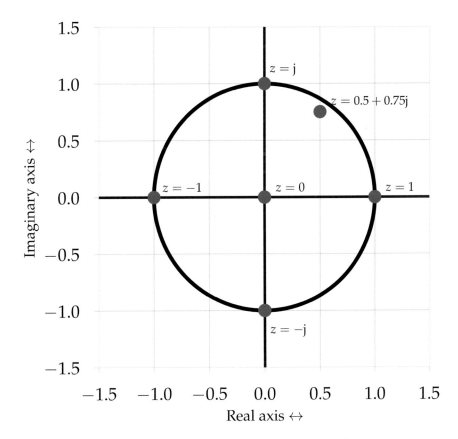

Figure 4.2 Complex numbers z represented as points in the plane.

When we visualize numbers in the **complex plane**, it is helpful to include a few extra pieces of information.

The horizontal axis is the *real number line*: numbers landing exactly on this line are purely real. Similarly, the vertical axis is the *imaginary number line*.

Finally, for reasons that will become clear below, it is also helpful to include a circle of radius 1. Numbers that land on the boundary of this circle all have equal distance from 0, and these numbers play a special role in signal processing.

It is tempting to think of a complex number as having two independent pieces (a and b), and it is sometimes helpful to do so. However, one must always keep in mind that this representation is primarily for convenience, and a complex number should always be treated holistically. A complex number has more structure than an arbitrary pair of points (x, y) because of the special interpretation of the imaginary unit j. We'll see why exactly this is the case soon.

4.2 BASIC OPERATIONS

4.2.1 Addition and subtraction

If we have two complex numbers $a+jb$ and $c+jd$, their sum is defined by independently summing the real parts $(a + c)$ and the imaginary parts $(b + d)$:

$$(a + jb) + (c + jd) = (a + c) + j(b + d).$$

Subtraction works similarly:

$$(a + jb) - (c + jd) = (a - c) + j(b - d).$$

In the complex plane, addition and subtraction can be thought of as displacement. If you imagine drawing an arrow from the origin 0 to the positions of z and w, then you can find the position of $z + w$ by picking up one of the arrows and moving its tail to coincide with the head of the other arrow. Subtraction works the same way, except that you would turn the arrow around 180° (see Figure 4.3).

4.2.2 Conjugation

Complex numbers have a new operation, which doesn't exist for other systems like reals and integers, called *conjugation*. The complex conjugate of a number $z = a+jb$, denoted as $\overline{z}$, is computed by negating *only the imaginary part*, and is denoted by an over-line:

$$\overline{z} = \overline{a + jb} = a - jb.$$

Visually, this operation reflects a point across the horizontal (real) axis in the complex plane, as seen below in Figure 4.4.

Conjugating a number twice reverts back to the original number (just like negating a number twice).

If a number is purely real (has $b = 0$), it lies exactly on the horizontal axis, and it is its own conjugate.

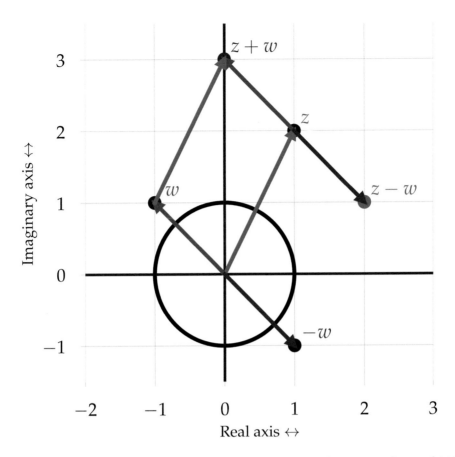

Figure 4.3 Addition or subtraction of complex numbers results in translation (shifting) in the plane.

4.2.3 Multiplication (take 1)

There are two ways to think about multiplication of complex numbers, depending on whether we use *rectangular* or *polar* form, which we'll see below. Both are equally valid, and produce the same results, but sometimes one may be more convenient than the other. Let's start with the rectangular form.

If we have two complex numbers $z = a + jb$ and $w = c + jd$, we can compute their product by using the rules of algebra and remembering that $j \cdot j = -1$:

$$
\begin{aligned}
z \cdot w &= (a + jb) \cdot (c + jd) && \text{Use definitions of } z, w \\
&= a \cdot c + a \cdot jd + jb \cdot c + jb \cdot jd && \text{FOIL multiply} \\
&= a \cdot c + j\,(a \cdot d + b \cdot c) + j^2\,(b \cdot d) && \text{Pull out imaginary units } j \\
&= a \cdot c + j\,(a \cdot d + b \cdot c) - b \cdot d && j^2 = -1 \\
&= (a \cdot c - b \cdot d) + j\,(a \cdot d + b \cdot c) && \text{Collect real and imaginary parts}
\end{aligned}
$$

This looks complicated – and it is! The key things to take away here are:

1. The product is well-defined, even if the formula is messy;

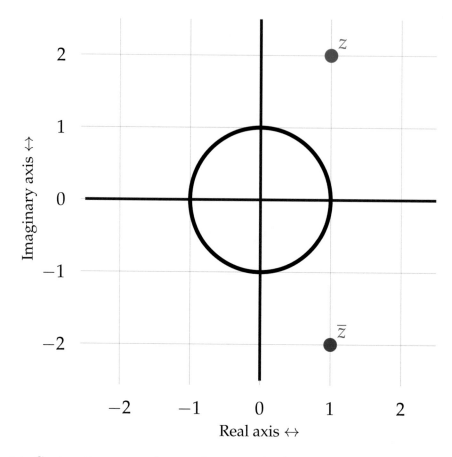

Figure 4.4 Conjugating a complex number $z \to \overline{z}$ reflects it across the horizontal axis (real number line).

2. When you multiply two complex numbers, their real and imaginary parts interact in a non-trivial way (see Figure 4.5);

3. When two purely imaginary numbers are multiplied together, in this case, $jb \cdot jd$, the result is *real*: $-b \cdot d$.

Example 4.1 (Multiplying complex by real). Let's keep $z = a + jb$ and multiply it by a purely real number x. If we follow the rules above, we get

$$z \cdot x = (a + jb) \cdot x$$
$$= a \cdot x + jb \cdot x$$

so the real x combines with both the real and imaginary parts of z.

Example 4.2 (Multiplying complex by imaginary). What if we multiply $z = a + jb$ a purely imaginary number jx? In this case, we get

$$z \cdot jx = (a + jb) \cdot jx$$
$$= a \cdot jx + jb \cdot jx$$
$$= -b \cdot x + ja \cdot x.$$

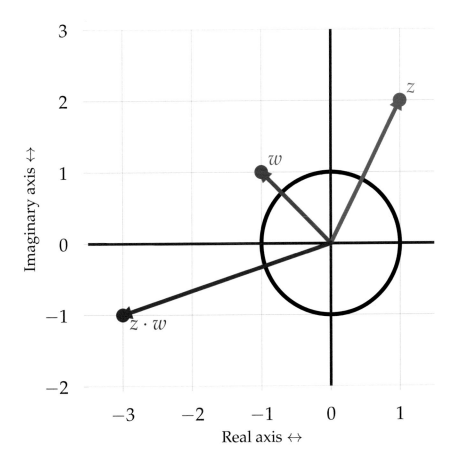

Figure 4.5 Complex multiplication implements a scaling and rotation in the complex plane.

In this case, the real and imaginary parts have exchanged places, and the new real part $(b \cdot x)$ has been negated.

Example 4.3 (Multiplying complex by its conjugate). Finally, what if we multiply a number $z = a + jb$ by its own conjugate $\bar{z} = a - jb$?

In this case, all of the imaginary components cancel each other out, and the resulting product is purely real (and non-negative):

$$
\begin{aligned}
z \cdot \bar{z} &= (a + jb) \cdot (a - jb) \\
&= a \cdot a + jb \cdot a - a \cdot jb - jb \cdot jb && \text{FOIL multiply} \\
&= a \cdot a + \cancel{jb \cdot a} - \cancel{a \cdot jb} - jb \cdot jb && \text{Cancel } -jab + jab = 0 \\
&= a \cdot a - jb \cdot jb && \text{Substitute } j^2 = -1 \\
&= a^2 + b^2
\end{aligned}
$$

We could say more about complex multiplication, but as we'll see, it's easier to think about if we use the *polar* form. To see how that works, we'll need to take a slight detour through the exponential function.

4.3 COMPLEX EXPONENTIALS

Recall the definition of the exponential function as an infinite summation:

> If you need a refresher on exponentials, see the *appendix*.

$$e^x = \sum_{n=0}^{\infty} \frac{x^n}{n!}$$

We normally think of this definition in terms of real-valued x, but the idea carries over to *complex* exponentials e^z for $z \in \mathbb{C}$ using exactly the same formula with z in place of x.

Because any complex z is the sum of its real and imaginary parts, it can be helpful to separate the exponential using the product rule:

$$e^z = e^{a+jb} = e^a \cdot e^{jb}$$

Since a is real, we already have a good handle on how e^a behaves. Let's focus on what happens just to that second factor, where the quantity in the exponent is purely imaginary: e^{jb}.

Fig. 4.6 shows what happens as we form better approximations to e^{jb} by including more terms in the summation.

4.3.1 Euler's formula

Fig. 4.6 shows that e^{jb} wraps imaginary numbers around the unit circle in the complex plane. By visual inspection, one can see that the point $b = 0$ maps to $z = 1 + 0j = 1$, which makes sense since $e^0 = 1$. More surprising are the points $b = \pm\pi$, which both map to $z = -1$.

This phenomenon is summarized succinctly by **Euler's formula**:

Note (Euler's formula). Let $\theta \in \mathbb{R}$ denote any angle (in radians). Then the *complex exponential* $e^{j\theta}$ is equal to a complex number with real and imaginary parts given as follows:

$$e^{j\theta} = \cos(\theta) + j \cdot \sin(\theta). \tag{4.1}$$

In plain language, this formula says that if you take the exponential of a purely imaginary number, the result will be a complex number at unit distance from the origin, and with an angle matching the number.

This is probably the most important formula in all of signal processing.

4.3.2 Polar and rectangular form

Euler's formula says what happens when you take the exponential of an imaginary number. This might seem like an odd thing to do, but let's take a step back to think about what this tells about exponentials of *complex* numbers $z = a + jb$.

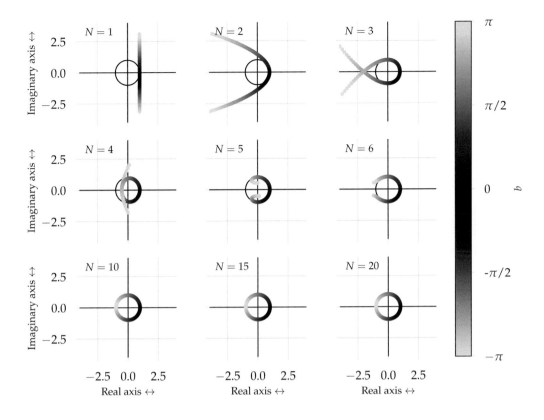

Figure 4.6 Partial summations illustrate how Euler's formula maps imagine numbers jb to the unit circle in the complex plane. Each plot illustrates the result of taking the first N terms of the summation $\sum_{n=0}^{N-1} (j \cdot b)^n / n!$ for values $-\pi \leq b \leq \pi$.

Remember that $e^z = e^{a+jb} = e^a \cdot e^{jb}$. Using Euler's formula, we can write this as

$$e^z = e^a \cdot e^{jb} = e^a \cdot (\cos(b) + j \cdot \sin(b))$$
$$= e^a \cdot \cos(b) + e^a \cdot j \cdot \sin(b).$$

What does this buy us?

Imagine picking any complex number $z = a + jb$. As we've seen earlier, this number can be represented by the point (a, b) in the complex plane. Equivalently, we can represent z in **polar** form in terms of its distance from the origin (which we call r) and the angle θ it makes with the real axis (see Figure 4.7).

Putting this picture back into equations, *any complex number z* can be equivalently expressed using Euler's formula as

$$z = r \cdot (\cos(\theta) + j \cdot \sin(\theta))$$
$$= r \cdot e^{j\theta}$$

for some radius $r \geq 0$ and angle θ. This is what we mean by the **polar form** of a complex number. The radius r of the circle is also called the **magnitude** of the complex number; the angle θ is called either the **angle** or the **phase** of the number.

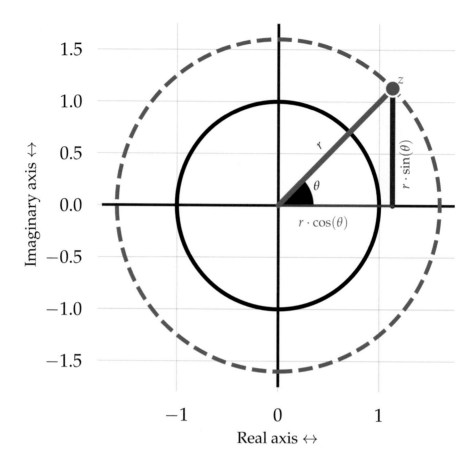

Figure 4.7 A complex number z can be expressed in polar form by finding its distance r from the origin, and the angle θ it makes with the real number line.

If we're given a complex number in polar form (that is, we're given (r, θ)), Euler's formula tells us how to convert it to rectangular form:

$$a = r \cdot \cos(\theta)$$
$$b = r \cdot \sin(\theta).$$

If instead, we're given a complex number in rectangular form (in terms of (a, b)), we can convert to polar form by using the Pythagorean theorem to find the radius, and some geometry to find the angle:

$$r^2 = a^2 + b^2 \qquad\qquad \Rightarrow r = \sqrt{a^2 + b^2}$$
$$\tan(\theta) = \frac{\sin(\theta)}{\cos(\theta)} = \frac{r \cdot \sin(\theta)}{r \cdot \cos(\theta)} = \frac{b}{a} \qquad\qquad \Rightarrow \theta = \tan^{-1}\left(\frac{b}{a}\right).$$

where $\tan^{-1}$ is the inverse or *arc-tangent* function.

Note (Computing angles in practice). In practice, you shouldn't compute the angle using the arc tangent directly. This is because dividing b/a can be numerically unstable when a is a small number.

```
# Don't do this.
np.arctan(1.0/0.0)
```

Instead, most mathematical software libraries give you functions to compute the inverse tangent without having to do the division first.

```
# Do this instead
np.arctan2(1.0, 0.0)
```

You can do even better by using the `numpy` functions to extract magnitude and angle from a complex number:

```
np.abs(0+1j)
np.angle(0+1j)
```

4.3.3 Summary

The key point here is that it's always possible to represent a complex number in either polar or rectangular form. Rectangular form is convenient when adding or subtracting complex numbers. As we'll see next, polar form is more convenient for multiplying and dividing.

4.4 MULTIPLICATION AND DIVISION

We previously saw how to multiply two complex numbers in rectangular form, but let's see what happens when we use polar form instead.

Say we're given two numbers $z = r \cdot e^{j\theta}$ and $w = s \cdot e^{j\phi}$. We can use the *rules of exponents* to simplify this:

$$
\begin{aligned}
z \cdot w &= r \cdot e^{j\theta} \cdot s \cdot e^{j\phi} \\
&= (r \cdot s) \cdot e^{j\theta} \cdot e^{j\phi} && \text{Collect exponentials together} \\
&= (r \cdot s) \cdot e^{j\theta + j\phi} && \text{Use rule: } e^x \cdot e^y = e^{x+y} \\
&= (r \cdot s) \cdot e^{j(\theta + \phi)} && \text{Pull out common factor of j.}
\end{aligned}
$$

The result in a new complex number in polar form, with magnitude $r \cdot s$ and angle $\theta + \phi$ as seen in figure 4.8. There's a helpful mnemonic device for remembering this:

Tip. Magnitudes multiply and **angles add**.

In the special case where $s = 1$, the magnitude is preserved, and all that happens is that the angle moves from $\theta \rightarrow \theta + \phi$. But adding angles is nothing more than **rotation!**

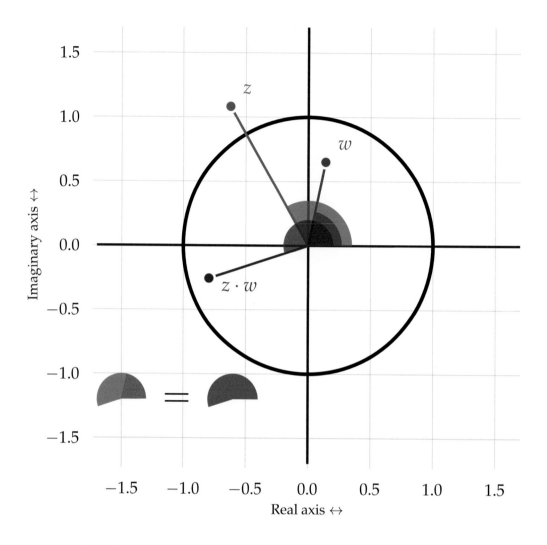

Figure 4.8 Complex multiplication can also be understood as adding the angles of the two numbers and multiplying their magnitudes.

4.4.1 Inversion (division)

We skipped inversion and division when working in rectangular coordinates. It's certainly not impossible to do these operations with rectangular coordinates, but it's much simpler in polar form if we use the rule

$$\frac{1}{e^x} = e^{-x}.$$

A complex number z can therefore be inverted to get z^{-1}:

$$\frac{1}{z} = \frac{1}{r \cdot e^{j\theta}} = \frac{1}{r} \cdot e^{-j\theta}.$$

That is, the radius is *inverted*, and the angle is *negated*.

Fig. 4.9 shows how a complex number and its inverse relate to each other. Some things to note:

- When z is outside the unit circle, the inverse is inside, and vice versa.

- If z is on the unit circle, so is z^{-1}, since $r = 1$ is its own inverse.

- There are two points at which z and z^{-1} coincide exactly: $z = 1$ and $z = -1$.

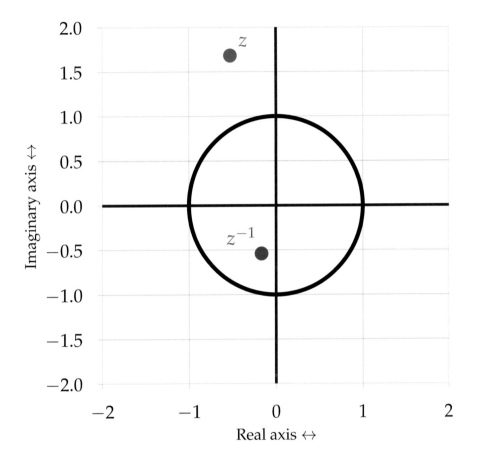

Figure 4.9 Inverting a complex number $z \to z^{-1}$ can be understood as inverting its magnitude and negating its angle.

If you want to divide two complex numbers $z = r \cdot e^{j\theta}$ and $w = s \cdot e^{j\phi}$ as z/w, it's equivalent to multiply by the inverse $z \cdot w^{-1}$. This results in

$$\frac{z}{w} = \frac{r \cdot e^{j\theta}}{s \cdot e^{j\phi}} = \frac{r}{s} \cdot e^{j(\theta-\phi)}$$

4.5 POWERS AND WAVES

Now that we've seen how complex numbers behave when multiplied together, we might also wonder what happens when a complex number is multiplied by itself repeatedly. Of specific importance in signal processing is the sequence

$$z^0, z^1, z^2, \cdots, z^n, \cdots$$

If we pick an arbitrary point in this sequence, z^n for some $n \geq 0$, we can express the corresponding term in polar form as follows

$$z^n = \left(r \cdot e^{j\theta}\right)^n = r^n \cdot e^{j \cdot n\theta},$$

where $z = r \cdot e^{j\theta}$ and the second equality follows from the product rule for exponents.

From this, we can see that the nth step in the sequence will place the point at angle $n \cdot \theta$, and radius r^n.

If $r < 1$, this sequence will have decreasing radius, and it will spiral into the origin. If $r > 1$, the sequence will have increasing radius, and it will spiral out. If $r = 1$, each element will have the same radius since $1^n = 1$: the sequence will therefore loop around the circle indefinitely.

If $0 \leq \theta \leq \pi$, the spiral will wind counter-clockwise, and otherwise, it will wind clockwise, as demonstrated in Figure 4.10. Larger angles (closer to $\pm\pi$) rotate faster than smaller angles (closer to 0).

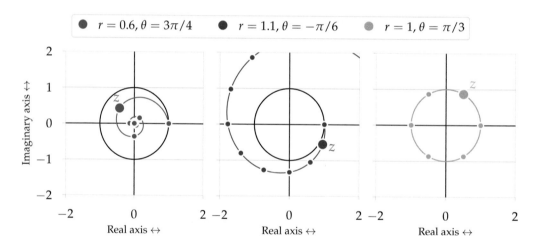

Figure 4.10 Raising a complex number z to a sequence of powers $n = 0, 1, 2, \ldots$ produces an inward spiral (if magnitude is less than 1); an outward spiral (if magnitude is greater than 1), or an orbit around the unit circle (if magnitude is exactly 1).

If we take just the real (or imaginary) part of z^n for the sequence of values of n, we'll see a sinusoid with decaying ($r < 1$), increasing ($r > 1$), or stable ($r = 1$) amplitude, as illustrated in Figure 4.11 for the real part (top) and imaginary part (bottom).

From these plots, we can see that any complex number z can be used to generate what is known as a **complex sinusoid** by raising it to successive powers z^n. This idea forms the basis of the (discrete) Fourier transform, which we will see in the next chapter.

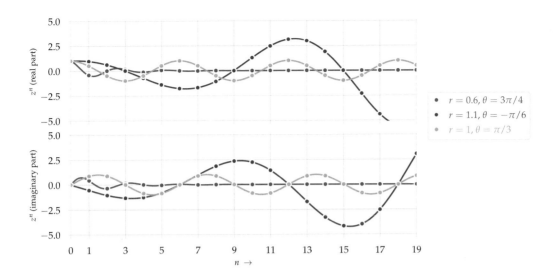

Figure 4.11 The real (top) and imaginary (bottom) components of z^n as n grows, for different choices of z.

4.6 EXERCISES

Exercise 4.1. How many complex numbers z are there, such that:

1. $z^2 = 1$?

2. $z^2 = -1$?

3. $z^3 = 1$?

Exercise 4.2. For each pair of numbers below, compute the sum $z + w$ and product $z \cdot w$. Check your results with Python code.

1. $z = j, w = 1/2 + j$

2. $z = 2 + 3j, w = 3 - 2j$

3. $z = e^{j \cdot \pi}, w = e^{-3j \cdot \pi/2}$

Exercise 4.3. What are the complex conjugates of the following numbers:

1. $z = 1 + 2j$?

2. $z = 3$?

3. $z = 3e^{j \cdot \pi}$?

4. $z = 3e^{j \cdot \pi/2}$?

Exercise 4.4. Let $z = e^{j \cdot \pi/3}$, and consider the sequence generated by raising it to successive powers: $z^0, z^1, z^2, \ldots, z^n, \ldots$

1. What is the smallest power $n > 1$ that satisfies $z^n = z$?

2. If we take $f_s = 100$, can you find a frequency $f > 0$ such that $\cos\left(2\pi \cdot f \cdot n/f_s\right) = \mathsf{Re}\left[z^n\right]$ (the real part of z^n) for all n?

## CHAPTER 5

The discrete Fourier transform

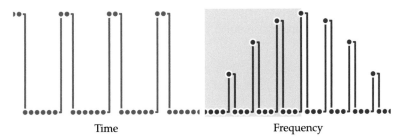

Time | Frequency

This chapter introduces the frequency domain representation of signals and the discrete Fourier transform. At a high level, the Fourier transform is just a different way of representing a signal: instead of using the sequence of sample values, we use combinations of sinusoids.

Converting between these two representations is done by measuring the *similarity* of a signal $x[n]$ to a collection of pre-defined sinusoids, resulting in a collection of similarity scores which characterize the signal. This process can be relatively complicated, so we'll introduce each step separately in this chapter:

1. Defining what we mean by *similarity* between signals

2. Comparing a signal to a collection of sinusoids (cosine transform)

3. Moving to complex sinusoids (the Fourier transform)

4. Understanding the frequency domain

5. Examples

From this chapter onward, we'll be making extensive use of *complex numbers*.

5.1 SIMILARITY

In this chapter, we'll develop the concepts of the **frequency domain** and the **discrete Fourier transform** from first principles. The underlying idea throughout is

DOI: 10.1201/9781003264859-5

that one can equivalently represent a signal by its sequence of sample values, or as a combination of sinusoids. The sequence of sample values, which we've been calling $x[n]$ is known as the **time-domain** representation because the variable n that we change to view the entire signal corresponds to time (sample index). The alternate representation that we'll develop in this chapter will instead vary the **frequency** of different sinusoids, and not depend explicitly on time.

To make this all work, we'll need to develop a way to convert from the time domain to the frequency domain. Our basic strategy will be to compare the input signal $x[n]$ to a collection of fixed reference signals. The result of these comparisons will be a collection of similarity measurements, which (loosely speaking) measure "how much" of each reference signal is in $x[n]$.

5.1.1 Measuring similarity

One could imagine many different ways to compare two signals $x[n]$ and $y[n]$.

The definition of **similarity** that we'll use is to go sample-by-sample, multiplying $x[n] \cdot y[n]$, and summing up the results to produce a single number S. In equations, this looks as follows:

$$S(x, y) = \sum_{n=0}^{N-1} x[n] \cdot y[n] \tag{5.1}$$

or equivalently in code:

```python
def similarity(x, y):
    '''Compute similarity between two signals x and y

    Parameters
    ----------
    x, y : np.ndarrays of equal length

    Returns
    -------
    S : real number
        The similarity between x and y
    '''

    # Initialize similarity to 0
    S = 0

    # Accumulate the sample-wise products
    N = len(x)
    for n in range(N):
        S = S + x[n] * y[n]

    return S
```

Fig. 5.1 provides a demonstration of this similarity calculation. Two signals $x[n]$ (a triangle wave, top plot) and $y[n]$ (a cosine wave, center plot) with $N = 32$ samples are compared by summing the sample-by-sample product ($x[n] \cdot y[n]$, bottom plot).

Each step of the loop (equivalently, term of the summation) is calculated independently, and the height of the bar on the right plot shows the running total. When the loop is finished (we've summed all samples $n = 0, 1, \ldots, 31$), we'll have computed the total similarity S.

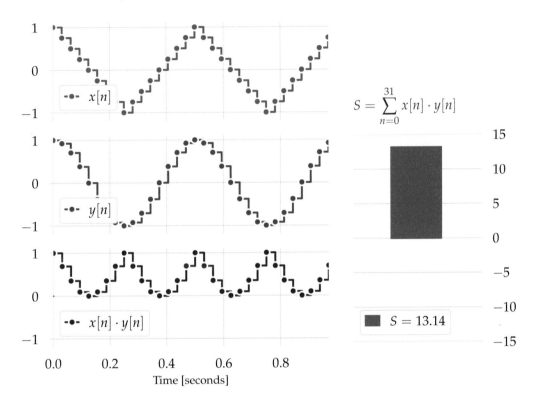

Figure 5.1 Similarity S between signals x and y is computed by summing the element-wise product of their samples $x[n] \cdot y[n]$.

This process might feel a bit like convolution, and the equations do look similar and involve many of the same ingredients. However, there are some key differences:

1. Convolution can be computed between any two signals. Similarity is only defined for two signals of exactly the same length.

2. Convolution involves shifting (delaying) one signal several times to produce the output. Similarity does not use shifting.

3. Convolution produces an output *signal* of multiple samples. Similarity produces a single number.

5.1.2 Interpreting similarity

Note that the values of $x[n]$ and $y[n]$ can be either positive or negative, and as a result, so can the similarity value $S(x, y)$. To build some intuition for how this similarity behaves, it can be helpful to imagine situations which could lead to different values of S: large positive, large negative, or near zero.

Whenever the signs of $x[n]$ and $y[n]$ agree, the contribution to the total sum S is positive. If this happens often (over many sample indices n), then S will consist primarily of positive contributions and therefore be a large positive number. This is the case in the example above, where the signs of the two signals are generally in agreement, leading to a high similarity score.

Likewise, when the signs disagree (e.g., $x[n] > 0 > y[n]$), the contribution to the sum is negative. Again, if this happens often, then S will consist primarily of negative contributions, and be a large negative number.

If neither of these situations occur, then the signs of $x[n]$ and $y[n]$ agree and disagree comparably often. The positive and negative contributions will be (approximately) balanced, so the total S will be near zero.

An example of each of these three cases is demonstrated in Fig. 5.2. It's important to remember that these cases are meant as qualitative examples, and the behavior of S will depend also on the *magnitude* of the sample values, not just their signs.

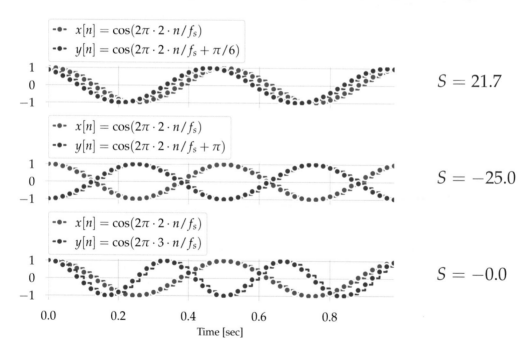

Figure 5.2 **Top**: two waves of the same frequency and small phase difference produce a high similarity score $S = 21.7$. **Middle**: two waves with maximal phase difference (*pi*) produce a minimal similarity score $S = -25.0$. **Bottom**: two waves of different frequencies can produce a similarity score of $S = 0$.

5.1.3 Summary

The definition of similarity that we've seen here can be applied to any pair of signals x and y, as long as they have the same length.

In the context of the Fourier transform, we're given a signal x, and the transform will be defined by the similarity scores produced by comparing x to a collection of signals. Our next step is to determine what that collection will be.

5.2 COMPARING TO SINUSOIDS

In this section, we'll begin to build up the foundations of the Fourier transform and the frequency domain. As stated in the introduction to this chapter, the entire idea of the Fourier transform is to represent arbitrary signals $x[n]$ as combinations of sinusoids:

$$
\begin{aligned}
x[n] = & A_0 \cdot \cos\left(2\pi \cdot f_0 \cdot \frac{n}{f_s} + \phi_0\right) \\
& + A_1 \cdot \cos\left(2\pi \cdot f_1 \cdot \frac{n}{f_s} + \phi_1\right) \\
& + \ldots
\end{aligned}
$$

Before we go any further with this idea, it is helpful to think carefully about what it would mean to represent an arbitrary discrete signal $x[n]$ in this way.

Warning. The representation that we develop in this section is **not** the Fourier transform.

This section will set up the foundations that we'll need to define the Fourier transform later in this chapter.

5.2.1 Band-limiting

Since $x[n]$ is a discrete signal taken at some sampling rate f_s, we will assume that the conditions of the Nyquist-Shannon theorem hold and that aliasing is not going to be a problem for us. More specifically, this means that we are assuming that the underlying continuous signal $x(t)$ which produced $x[n]$ is **band-limited** to the frequency range $\pm\frac{f_s}{2}$. This implies that when representing $x[n]$, we will not need to use frequencies outside of this range.

That said, there are still infinitely many frequencies in the range $[-f_s/2, f_s/2]$ (e.g., 1 Hz, 1/2 Hz, 1/4 Hz, ... and we can't compare to all of them and still have a finite representation of the signal. We'll need to be careful in choosing which frequencies we compare to.

5.2.2 Periodicity

Recall back in *chapter 1* that periodic signals must repeat forever, and this includes the sinusoids that will form the basis of any frequency-domain representation. This also carries through to (finite) combinations of sinusoids. Any signal that we try to represent as a combination of sinusoids, therefore, must also be assumed to be periodic.

Finite vs infinite?

Things get weird when we allow for infinitely many sinusoids.

Fortunately, computers (like ourselves) are finite in nature, so let's not worry about it.

This presents a bit of a problem for us, as up to now, we've been assuming that signals are silent for all negative time: $x[n] = 0$ for $n < 0$. A (non-empty) signal with an infinitely long stretch of silence cannot be periodic, so what can we do?

To make progress here, we'll need to change our assumptions. Rather than assume that signals are silent for all time before recording, and we'll assume that a finite signal $x[0], x[1], \ldots, x[N-1]$ represents one cycle of a loop that repeats forever as shown in Figure 5.3. Under this assumption, once we've observed these N samples, there is nothing new to learn from further observation, since the repetition assumption gives us for any sample index n:

$$x[n] = x[n+N] = x[n+2 \cdot N] = \cdots = x[n+k \cdot N] \qquad \text{for any } k \in \mathbb{Z}. \qquad (5.2)$$

This also gives a new view to negative sample indices as offsets from the end of the signal:

$$x[-n] = x[-n+N] = x[N-n]. \qquad (5.3)$$

This assumption is demonstrated visually below, where a signal $x[n]$ of duration 1.5 seconds (sampled at $f_s = 16$, so $N = 24$) is shown along with its implied repetitions. These repetitions are implied to go on forever in each direction of time. Note that the repeating signal need not be continuous at the repetition boundaries: all that matters is that the observed signal repeats exactly.

Array indexing

More generally, we can assume

$$x[n] = x[n \mod N]$$

where $n \mod N$ is the remainder of dividing n by N.

In Python, this is done by saying

```
x[n % N]
```

Python also allows us to use negative array indices (up to -N) directly to index from the end of the array:

```
x[-1]   # Last sample
```

5.2.3 Choosing reference frequencies

As mentioned above, even though we're assuming $x[n]$ represents a band-limited signal, there are still infinitely many frequencies within the band. Our periodicity

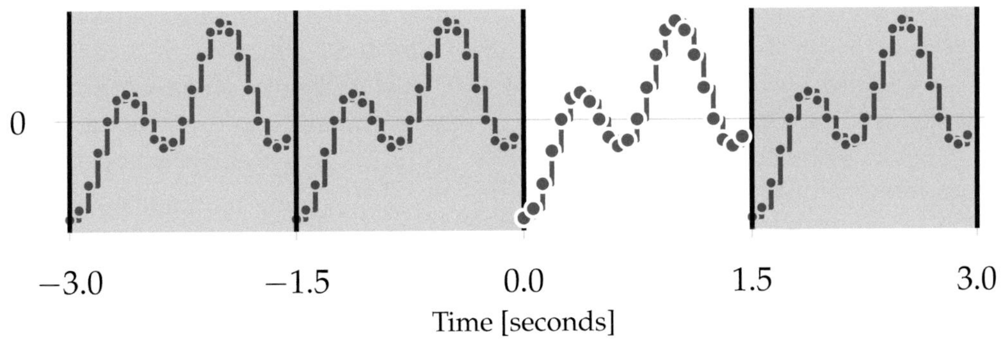

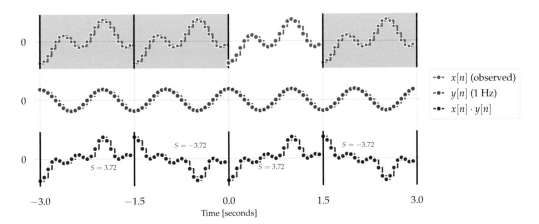

Figure 5.3 A signal $x[n]$ of some finite number of samples N (highlighted region) is assumed to repeat forever. Vertical lines denote the boundaries of each repetition of x, additional repetitions are depicted in the shaded region.

assumption will help us narrow this down to a finite set that we can use to form our reference signals.

In Figure 5.3, we see several identical repetitions of the observed signal $x[n]$. Imagine comparing the full repeating signal to a sinusoid of some frequency f, and computing the *similarity score* independently for each individual repetition. Ideally, it should not matter which repetition of x we look at: they should all produce the same similarity score. Unfortunately, this is not the case.

For example, if we take $f = 1$ [Hz] to get a reference signal

$$y[n] = \cos\left(2\pi \cdot \frac{n}{f_s}\right)$$

we would observe Figure 5.4:

Figure 5.4 Top: a repeating signal x, with one observed repetition highlighted. Middle: a reference sinusoid y. Bottom: each repetition of x produces a different similarity score when compared to the corresponding segment of the reference signal y.

Each repetition of the signal $x[n]$ produces a different similarity score when compared to the 1 Hz reference signal $y[n]$, alternating between $S = +3.72$ and $S = -3.72$. As a result, it's unclear how we should define the similarity between x and y: if it's the similarity between one repetition, which one do we choose? Or do we combine scores from multiple repetitions, and if so, how would that work if they go on forever? In this example, we only had two distinct score values to deal with, but it's possible to construct examples where there are many more than that.

This problem arises because the period of the reference signal (1 second) does not line up with the period x, so the reference signal appears to start at a different point within each repetition. In the figure, we observe the reference signal starting either high (1) or low (-1) in alternating repetitions of x.

If we can ensure that the reference signal always starts at the same position in each repetition of x, then the problem vanishes: each repetition will produce the same similarity score, so we only need to examine one of them. This observation leads us to the concept of an *analysis frequency*.

Definition 5.1 (Analysis frequency). For a sampling rate f_s and a signal length N, an **analysis frequency** f_m is any frequency which exactly completes a whole (integer) number of cycles m in N samples at f_s.

Equivalently:

$$ f_m = \frac{m}{N} \cdot f_s \left[\frac{\text{cycles}}{\text{sec.}} \right] \qquad\qquad m = 0, 1, 2, \ldots $$

Aside: bins and bands

We use the term **analysis frequency** to distinguish a special set of frequencies determined by the sampling rate f_s and number of samples N. As we'll see later in this chapter, these frequencies play a critical role in the construction of the discrete Fourier transform.

In other places (books, software implementations, etc), you may see other terms used for this idea, such as *frequency bins* or *frequency bands*. The reason behind that choice of terminology should become clear in the next chapter when we cover *spectral leakage*, but throughout the text, we'll stick to the more specific terminology of analysis frequencies.

Just note that these terms are often used interchangeably, so if you see reference to a "frequency bin" in the context of a Fourier transform, it is most likely describing either an analysis frequency (or a quantity derived from one, such as a similarity score).

Since we will be sampling waves at these frequencies at rate f_s over the full signal duration N, it is sometimes more useful to think of analysis frequencies in units of `[cycles/signal-duration]`, or in the notation above, the ratio m/N.

Analysis frequencies have exactly the characteristic that we want: a shift by N samples will leave the signal unchanged, so each repetition of x is guaranteed to produce the same similarity score. For example, if we had chosen $m = 1$ (one cycle

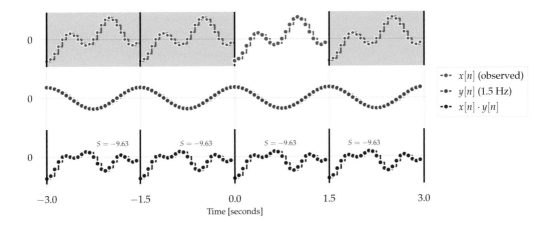

Figure 5.5 If the reference sinusoid corresponds to an analysis frequency, each repetition of x will produce the same similarity score.

per signal duration) so that $f = 1/N \cdot f_s$ (1.5 Hz for this case), we'd observe the result in Figure 5.5.

Since we are now guaranteed that each repetition produces the same similarity score, we only need to look at one of them.

One is much less than infinity, so we're making progress.

How many distinct analysis frequencies are there?

Now that we've settled on *which* frequencies we'll be comparing our signal to, the next natural question is **how many analysis frequencies are there**? Since each analysis frequency will give us a different similarity score, the number of frequencies determines the size of the representation of our signal. Our goal will be to find a minimal set of analysis frequencies which can accurately represent the signal.

As defined above, each analysis frequency is determined by a non-negative integer m which counts how many full cycles (positive or negative) are completed over the duration N samples. There are, of course, infinitely many values for m, but let's see what happens.

For an index m, we have analysis frequency $f_m = m/N \cdot f_s$, and the samples of the corresponding reference signal y_m (using cosine waves) will be determined by:

$$y_m[n] = \cos\left(2\pi \cdot \frac{m}{N} \cdot f_s \cdot \frac{n}{f_s}\right) = \cos\left(2\pi \cdot \frac{m}{N} \cdot n\right). \tag{5.5}$$

Figure 5.6 illustrate the sampled reference signals for the first few analysis frequencies with $f_s = 20$ and $N = 30$ (1.5 seconds).

If we continue this process a little further ($m = \ldots, N-2, N-1, N, N+1, \ldots$), what happens?

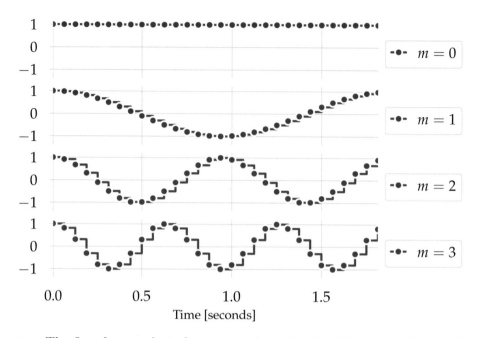

Figure 5.6 The first four analysis frequencies for a signal of duration 1.5 seconds.

Let's take $m = N + 1$ as an example. In this case, we have

$$
\begin{aligned}
f_{N+1} &= \frac{N+1}{N} \cdot f_s \\
&= \frac{N}{N} \cdot f_s + \frac{1}{N} \cdot f_s \\
&= f_s + \frac{1}{N} \cdot f_s \\
&= f_s + f_1.
\end{aligned}
$$

This is our old friend, *aliasing*. Analysis frequency $m = N + 1$ gives us exactly the same samples as $m = 1$ because they are related by the bandwidth of the signal f_s. Comparing our signal x to the reference signal for $m = N + 1$ would therefore give us exactly the same similarity score as if we had compared to the $m = 1$ signal. So let's not bother with it.

More generally, for any $m \geq N$, the corresponding frequency will always have an alias given by $m \mod N$. This answers our question: for a signal of N samples, there are only N such frequencies ($m = 0, 1, 2, \ldots, N - 2, N - 1$) which produce distinct reference signals. How fortunate that this is exactly the same number as the length of our signal.

A closer look at analysis frequencies

Now that we have our analysis frequencies chosen, it's worth taking a closer look to make sure we really understand what's going on. The plot above shows the first few ($m = 0, 1, 2, 3$), but what happens for larger values?

Recall that we've assumed the signal $x[n]$ to be band-limited to satisfy the Nyquist-Shannon theorem, so it cannot contain energy from frequencies above $f_s/2$ (the Nyquist frequency). The Nyquist frequency appears in our analysis frequencies (when N is even, anyway) as $m = N/2$:

$$f_{N/2} = \frac{N}{2N} \cdot f_s = \frac{f_s}{2},$$

but m itself can range all the way up to $N - 1 > N/2$.

Again, aliasing can help us make sense of this. When $N - 1 \geq m > N/2$, the corresponding frequency will lie outside the band limits of the signal. However, it will have an alias within the band, which we can find by subtracting f_s:

$$f_m - f_s = \frac{m}{N} \cdot f_s - f_s$$

$$= \frac{m}{N} \cdot f_s - \frac{N}{N} \cdot f_s$$

$$= \frac{m - N}{N} \cdot f_s.$$

Since $m < N$, this will be a negative frequency, *but that's okay.* The important thing is that it will be within the range $[-f_s/2, +f_s/2]$.

Tip. Analysis frequencies for $m \geq N/2$ correspond to **negative frequencies**.

So far, in this section, we've been assuming that the reference signals are produced by cosine waves, for which negative frequencies are indistinguishable from positive frequencies. (Remember, $\cos(\theta) = \cos(-\theta)$.) This means that the distinction isn't so important right now. However, it will become important in the full context of the Fourier transform, which we will see in the following sections.

5.2.4 Bringing it all together

To finish off this section, let's combine everything we've seen so far, and finally compare a signal $x[n]$ to a full collection of reference signals. For each analysis frequency f_m, we'll compute a similarity score S_m, and the collection of scores $[S_0, S_1, \ldots, S_{N-1}]$ will form the representation of our signal. We'll use the same example signal as above, but we'll now focus on a single repetition of $x[n]$ rather than the infinitely repeating sequence.

Figure 5.7 shows the signal x, reference signals y_m for $m = 0, 1, 2, \ldots, 9$ in the left column. The middle column shows the product of $x[n] \cdot y_m[n]$ for each reference signal y_m, and the right column shows the resulting similarity scores.

Showing all $N = 150$ comparisons in the form above would take quite a bit of space, so we only show the first ten to illustrate the idea.

More frequently, we omit the center column, and go directly to the third column. For our signal, if we show all N similarity scores in one plot, it looks like Figure 5.8. This kind of plot is called a **spectrum**. By analogy to colors, you can think of the length of each bar as measuring "how much" of each analysis frequency is present in the signal $x[n]$. You may have encountered something like this previously, for example, in a music player application, digital audio workstation, or on a stereo system.

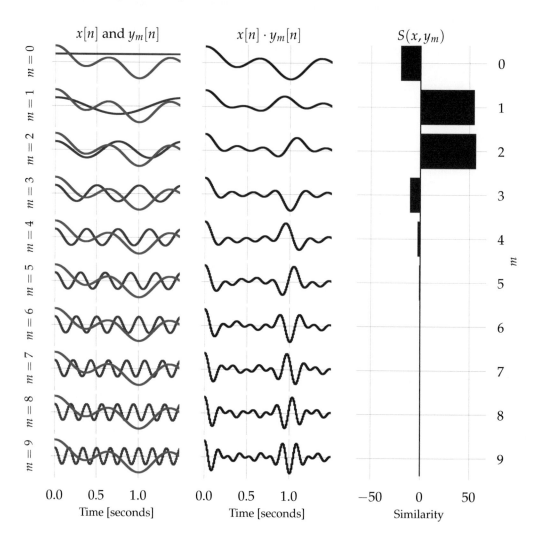

Figure 5.7 **Left**: a signal x overlaid with the first 10 reference sinusoids y_m. **Middle**: the per-sample product. **Right**: the resulting similarity scores for each reference sinusoid.

We'll have much more to say about frequency spectra in the following sections, but for now we'll leave it here.

5.2.5 Summary

As mentioned at the top of this section, the process that we've just seen is **not quite** the Fourier transform, but we're almost there.

The key things to remember from this section:

1. We now assume that a signal $x[n]$ of N samples represents one observation of an infinitely repeating loop.

2. Analysis frequencies complete a whole number of cycles in N samples.

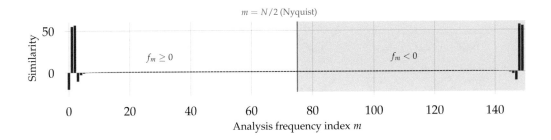

Figure 5.8 A bar-plot illustrating the similarity score for the input signal compared to the full set of analysis frequencies. Indices $m = 0, 1, \ldots, N/2$ correspond to positive frequencies, and $m \geq N/2$ (shaded region) correspond to negative frequencies.

3. We can represent a signal x according in terms of its similarity to sinusoids at our chosen analysis frequencies.

5.3 ANALYSIS FREQUENCIES

In the previous section, we saw how to setup comparisons between an input signal $x[n]$ and a collection of reference signals y_m defined to be sinusoids (cosine waves) with analysis frequencies determined by the length N of the signal (and the sampling rate f_s).

In this section, we'll dig a little deeper, and see how this kind of comparison behaves for specific choices of x. This will both solidify our understanding of the process, and also illustrate some short-comings of the method that we'll need to improve.

5.3.1 Comparing analysis frequencies

Let's see what happens when $x[n]$ is also a sinusoid at one of our chosen analysis frequencies.

For this example, we'll take $f_s = 20$, $N = 40$ (2 seconds), and take x to be the analysis frequency for $m = 3$ [cycles/signal-duration]. Since the signal duration is 2 seconds, this corresponds to a frequency of 3/2 [Hz]. Our sampled wave will be given by

$$x[n] = \cos\left(2\pi \cdot \frac{3}{40} \cdot n\right). \tag{5.6}$$

Fig. 5.9 shows the similarity comparisons and resulting scores for the first five reference signals. In this case, all scores are 0 except for $m = 3$, which corresponds exactly to the frequency of our test signal $x[n]$.

Why does this happen?

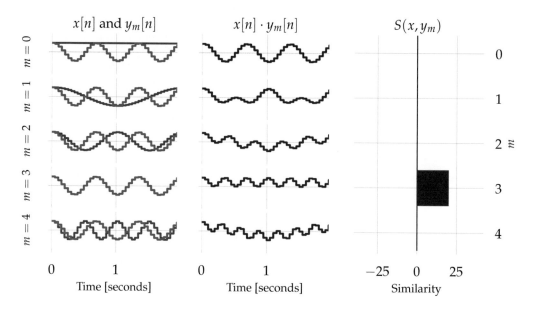

Figure 5.9 Comparing a sinusoid $x[n]$ at an analysis frequency to each reference signal $y_m[n]$.

Understanding analysis frequencies

Our definitions of *similarity* (5.1) and *reference signal* (5.5) lead to the following equation when $x[n]$ is the cosine wave given above in (5.6):

$$
\begin{aligned}
S(x, y_m) &= \sum_{n=0}^{N-1} x[n] \cdot y_m[n] \\
&= \sum_{n=0}^{N-1} \cos\left(2\pi \cdot \frac{3}{N} \cdot n\right) \cdot \cos\left(2\pi \cdot \frac{m}{N} \cdot n\right)
\end{aligned}
\tag{5.7}
$$

We can simplify (5.7) with a bit of trigonometry. The **product-to-sum** rule for cosines gives us for any angles θ and ϕ:

$$
\cos(\theta) \cdot \cos(\phi) = \frac{1}{2}\left(\cos(\theta + \phi) + \cos(\theta - \phi)\right)
$$

If for each sample index n, we let

$$
\theta = 2\pi \cdot \frac{3}{N} \cdot n
$$

be the angle of the first factor in the summation (the x part), and let

$$
\phi = 2\pi \cdot \frac{m}{N} \cdot n
$$

be the angle of the second factor (the y part), we would get

$$\theta + \phi = 2\pi \cdot \frac{3 + m}{N} \cdot n$$

$$\theta - \phi = 2\pi \cdot \frac{3 - m}{N} \cdot n,$$

and each term of the summation in (5.7) can be rewritten as:

$$\cos\left(2\pi \cdot \frac{3}{N} \cdot n\right) \cdot \cos\left(2\pi \cdot \frac{m}{N} \cdot n\right) = \frac{1}{2} \cdot \left(\cos\left(2\pi \cdot \frac{3 + m}{N} \cdot n\right)\right.$$

$$\left. + \cos\left(2\pi \cdot \frac{3 - m}{N} \cdot n\right)\right)$$

$$= \frac{1}{2}\left(y_{3+m}[n] + y_{3-m}[n]\right).$$

This tells us that in this case, S can be computed by averaging the samples of reference signals y_{3+m} and y_{3-m}.

More generally, if x was the k'th analysis frequency, we would have y_{k+m} and y_{k-m}.

Plugging this back into Equation (5.7), we get

$$S(x, y_m) = \sum_{n=0}^{N-1} \frac{1}{2}\left(y_{3+m}[n] + y_{3-m}[n]\right)$$

$$= \frac{1}{2}\left(\sum_{n=0}^{N-1} y_{3+m}[n]\right) + \frac{1}{2}\left(\sum_{n=0}^{N-1} y_{3-m}[n]\right)$$

where the second step follows by breaking the summation into pieces only concerned with y_{3+m} or y_{3-m}.

Now, let's consider two cases. If $m \neq 3$, then both of these signals will have non-zero frequency. Because they are both analysis frequencies, they will complete a whole number of cycles in the duration of the signal. Though we haven't proven it yet, this means that the sample values will *sum to zero*. (We'll prove this fact and a slightly more powerful theorem a few sections from now.)

Since both summations will sum to zero, the full similarity score $S(x, y_m) = 0$.

On the other hand, if $m = 3$, then our two signals will be $y_{3+3} = y_6$ and $y_{3-3} = y_0$. The first signal y_6 will total to zero by the argument above. The second signal, y_0 however will *not* total to zero, because $y_0[n] = 1$ for all n. As a result,

$$S(x, y_3) = \frac{1}{2}\sum_{n=0}^{N-1} y_0[n] = \frac{1}{2} \cdot N.$$

There is a third case that can happen (but not in the example above), where x is generated by either the highest analysis frequency $k = N/2$ (the Nyquist frequency) or if $k = 0$ (constant signal). If this happens, and $m = k$, then

$$y_{k+m} = y_{2 \cdot N/2} = y_N = y_0 \qquad\qquad \text{if } k = m = \frac{N}{2}$$

$$y_{k+m} = y_{0+0} = y_0 \qquad\qquad \text{if } k = m = 0.$$

In either event, both terms in the summation become $N/2$ and the total score is N.

This explains the observation above: similarity between analysis frequencies is either 0, $\frac{N}{2}$, or N, and the latter cases only occur when the reference signal y_m has the same frequency as the input x.

Warning. This cancellation **only** happens if x is a wave at an analysis frequency. Waves at non-analysis frequencies will not generally share this property.

5.3.2   Non-analysis frequencies

If instead of an analysis frequency, we had chosen x to be a sinusoid at a frequency that **does not** complete a whole number of cycles over the signal duration, we could still try to apply the reasoning above, but it's not going to work out.

For example, if we take $m = 1.5$ [cycles/signal-duration] – which is not an analysis frequency because 1.5 is not an integer – we would have a wave $x[n]$ at $1.5/2 = 3/4$ [Hz]:

$$x[n] = \cos\left(2\pi \cdot \frac{1.5}{40} \cdot n\right).$$

Like before, Fig. 5.10 shows the similarity comparisons and resulting scores for the first five reference signals. In this case, the scores are all non-zero because the wave doesn't precisely line up with itself at N samples.

This phenomenon is known as **spectral leakage**: a wave at a non-analysis frequency will *leak* across the entire frequency range, and appear to be a little similar to each analysis frequency.

We'll have more to say about this in subsequent chapters, but for now, it's important to understand that waves of different frequencies can still produce a non-zero similarity score.

5.4   PHASE

The examples we've seen so far have all used cosine waves, but what if we had used a sine wave instead of a cosine for our signal x?

Continuing the example in the previous section, we'll use $m = 3$ to generate a sine wave at an analysis frequency with $f_s = 20$ and $N = 40$:

$$x[n] = \sin\left(2\pi \cdot \frac{3}{40} \cdot n\right).$$

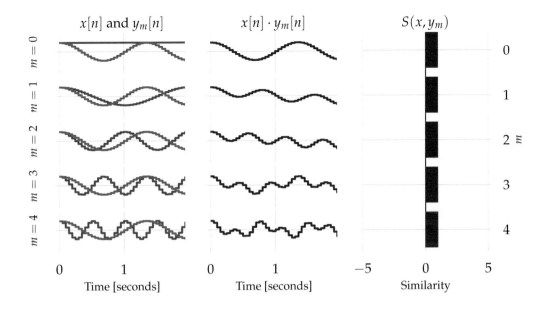

Figure 5.10 Comparing a sinusoid $x[n]$ at a non-analysis frequency to each reference signal y_m results in non-zero similarities S.

Fig. 5.11 is not an error – all similarity scores here are 0, **including** the comparison to $m = 3$.

To make sense of this, we'll first convert the sine wave into a standard cosine form by using Equation (1.5):

$$x[n] = \sin\left(2\pi \cdot \frac{3}{N} \cdot n\right) \qquad \text{by definition of } x$$

$$= \cos\left(\frac{\pi}{2} - 2\pi \cdot \frac{3}{N} \cdot n\right) \qquad \text{by the conversion rule}$$

$$= \cos\left(2\pi \cdot \frac{3}{N} \cdot n - \frac{\pi}{2}\right) \qquad \cos(\theta) = \cos(-\theta).$$

Now that we have x in cosine form, we can apply the reasoning from above to calculate each similarity score:

$$S(x, y_m) = \sum_{n=0}^{N-1} \cos\left(2\pi \cdot \frac{3}{N} \cdot n - \frac{\pi}{2}\right) \cdot \cos\left(2\pi \cdot \frac{m}{N} \cdot n\right)$$

$$= \frac{1}{2} \sum_{n=0}^{N-1} \left(\cos\left(2\pi \cdot \frac{3+m}{N} \cdot n - \frac{\pi}{2}\right) + \cos\left(2\pi \cdot \frac{3-m}{N} \cdot n - \frac{\pi}{2}\right) \right)$$

This looks almost identical to our first example, except that we now have phase offsets of $-\pi/2$ in both terms of the summation. If $m \neq 3$, this phase difference will not matter: summing over all samples $n = 0 \ldots N - 1$ will still produce a total of $S = 0$.

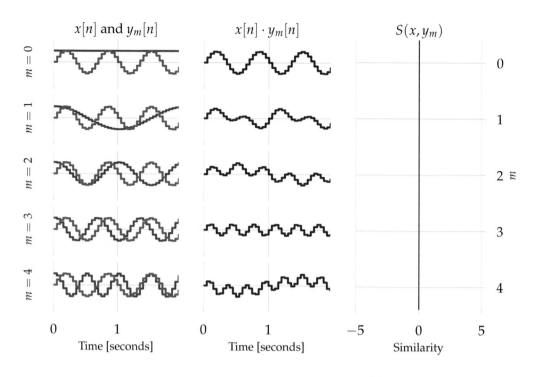

Figure 5.11 A sine wave $x[n]$ at an analysis frequency results in similarities $S = 0$ for all reference signals y_m.

However, if $m = 3$, the first term (frequency index $3 + m$) will again cancel to 0 when summed, but the second term ($3 - m = 0$) will simplify to

$$\cos\left(2\pi \cdot \frac{0}{N} \cdot n - \frac{\pi}{2}\right) = \cos\left(-\frac{\pi}{2}\right) = \cos\left(\frac{\pi}{2}\right) = 0.$$

So, rather than getting a contribution of $\cos(0) = 1$ for each term in the summation like our first example, we instead get a contribution of 0, and the total summation results in $S = 0$.

This is a huge problem.

Remember that our goal is to represent the frequency content of a signal x by comparing it against a collection of reference signals of known frequencies. But what we've just shown is that a signal can have exactly the same frequency as one of our reference signals, **and still produce a score of 0**.

This example of a sine wave is in some ways the worst-case scenario. In a bit more generality, we can consider a signal x with an analysis frequency index $m \neq 0$ and arbitrary phase offset ϕ:

$$x[n] = \cos\left(2\pi \cdot \frac{m}{N} \cdot n + \phi\right)$$

and by substituting ϕ for $-\pi/2$ in the above derivation, we generally have

$$S(x, y_m) = \begin{cases} \frac{N}{2} \cdot \cos\phi & \text{if } m \neq \frac{N}{2} \\ N \cdot \cos\phi & \text{if } m = \frac{N}{2}. \end{cases} \tag{5.8}$$

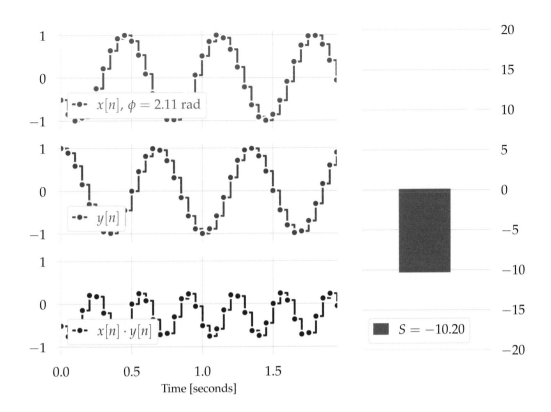

Figure 5.12 x (top) is a sinusoid of the same analysis frequency as the reference signal y (middle), but with non-zero phase ϕ. The sample-wise product $x \cdot y$ (bottom) and similarity score S (right) both depend on the phase ϕ.

5.4.1 Summary

By now, we've seen a few interesting quirks that arise when we compare sinusoids. To quickly recap:

1. If x is **cosine wave** at an analysis frequency, it will have non-zero similarity only for reference signal of the same frequency.

2. If x is a cosine wave at a **non-analysis** frequency, it will have non-zero similarity to all reference signals.

3. If x is a **sine wave** at an analysis frequency, it will have zero similarity even if the reference signal is of the same frequency.

4. If x is a sinusoid of arbitrary phase at an analysis frequency, its similarity scores will generally depend on its phase.

In the next section, we'll see how to resolve the phase issue to provide a more robust representation: *the Fourier transform*.

5.5 THE DISCRETE FOURIER TRANSFORM

In this section, we will formally introduce the discrete Fourier transform (DFT). The DFT is the *digital* version of the Fourier transform, which was originally developed by Joseph Fourier in the early 19th century as a way to model heat flow with differential equations [Fou22]. The Fourier transform proved to be useful for many different applications, especially in audio and signal processing.

Many texts introduce the DFT by first starting with the continuous Fourier transform (using integral calculus), and then show the DFT as a discrete-time approximation of the continuous Fourier transform. We'll take a different approach here, motivating the DFT from first principles entirely in the discrete time domain.

Aside: the Fast Fourier Transform

If you have previously come across the term *Fast Fourier Transform* (FFT), you might be wondering how the DFT relates to the FFT.

We'll cover the FFT later in *Fast Fourier Transform*, but the key take-away is that the FFT is an *algorithm* for efficiently computing the *DFT*. It is not a separate transformation.

5.5.1 Dealing with phase

Recall that in the previous section, when we compare a sinusoid $x[n]$ to a collection of reference signals

$$y_m[n] = \cos\left(2\pi \cdot \frac{m}{N} \cdot n\right),$$

the resulting similarity score $S(x, y_m)$ depends on the phase of the signal, and whether or not the frequency of x is an analysis frequency.

In the worst case, if

$$x[n] = \sin\left(2\pi \cdot \frac{m}{N} \cdot n\right) = \cos\left(2\pi \cdot \frac{m}{N} \cdot n - \frac{\pi}{2}\right),$$

then the similarity score will be 0 even if the frequency matches one of our reference signals. This representation therefore cannot distinguish an out-of-phase sinusoid from a null signal $x[n] = 0$.

Comparing to multiple references

In the specific case above, we could try to resolve the problem by changing the reference signals y to use sin instead of cos. This will fix the problem if x is a sin wave, but then it will fail in exactly the same way if x is a cos wave.

But what if we try both?

Let's see what happens if we take x as a sinusoid at an analysis frequency with arbitrary phase, and compare it to two reference signals at the same frequency: one is cos and one is sin.

In the following example, we'll continue with the settings from the previous section: $f_s = 20$ [samples / sec], $N = 40$ [samples], and $m = 3$ (equivalently, $f_0 = 1.5$ [cycles/sec]). This gives an input signal

$$x[n] = \cos\left(2\pi \cdot \frac{m}{N} \cdot n + \phi\right) \tag{5.9}$$

where ϕ is allowed to vary.

We'll compare to reference signals

$$\cos\left(2\pi \cdot \frac{m}{N} \cdot n\right), \qquad \text{and} \qquad \sin\left(2\pi \cdot \frac{m}{N} \cdot n\right)$$

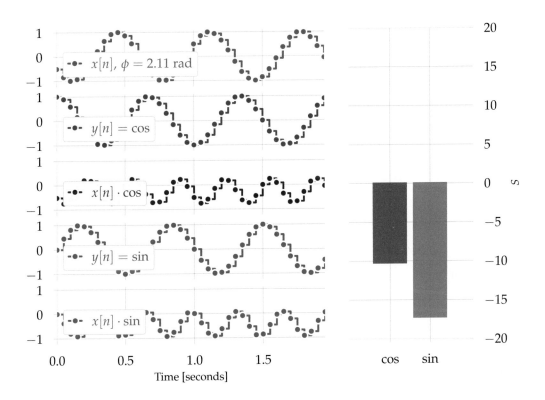

Figure 5.13 A sinusoid x at an analysis frequency but with non-zero phase ϕ (top plot) is compared to two reference signals: cos (second plot) and sin (fourth plot), producing element-wise products (third and fifth plots). On the right, the resulting similarity scores for both reference signals are shown.

Fig. 5.13 illustrates what happens to similarity scores as we vary the phase ϕ of x while comparing to both cosine and sine waves. Note that although either score can be positive, negative, or 0, **they cannot both be 0 simultaneously**.

Recall from (5.8) that the similarity score for x and the reference signal $\cos(2\pi \cdot n \cdot m/N)$ depends on ϕ: $S = \frac{N}{2} \cdot \cos\phi$ (if $m \neq N/2$). By similar reasoning, we can calculate the similarity between x and the sin reference signal:

$$
\begin{aligned}
S &= \sum_{n=0}^{N-1} \cos\left(2\pi \cdot \frac{m}{N} \cdot n + \phi\right) \cdot \sin\left(2\pi \cdot \frac{m}{N} \cdot n\right) \\
&= \sum_{n=0}^{N-1} \left(\frac{1}{2} \cdot \sin\left(2\pi \cdot \frac{m+m}{N} \cdot n + \phi\right)\right. \\
&\qquad\qquad \left. - \frac{1}{2} \cdot \sin\left(2\pi \cdot \frac{m-m}{N} \cdot n + \phi\right)\right) \qquad \text{product-to-sum rule} \\
&= \sum_{n=0}^{N-1} \frac{1}{2} \cdot \sin\left(2\pi \cdot \frac{2m}{N} \cdot n + \phi\right) - \frac{1}{2} \cdot \sin(\phi) \qquad \text{cancel } m - m = 0 \\
&= \left(\frac{1}{2} \sum_{n=0}^{N-1} \sin\left(2\pi \cdot \frac{2m}{N} \cdot n + \phi\right)\right) - \frac{N}{2} \cdot \sin(\phi) \quad \text{pull} -\frac{1}{2}\sin(\phi) \text{ out of sum}
\end{aligned}
$$

If $m \notin \{0, N/2\}$, then the first term of the summation will total to 0 because it completes a whole number of cycles in N samples. The resulting score in this case is

$$
S = -\frac{N}{2} \cdot \sin(\phi) \qquad \text{if } m \notin \{0, N/2\}.
$$

Otherwise, if $m = 0$ or $m = N/2$, the first term simplifies to

$$
\sin\left(2\pi \cdot \frac{2m}{N} \cdot n + \phi\right) = \sin\left(2\pi \cdot \frac{0}{N} \cdot n + \phi\right) = \sin(\phi).
$$

Combining this with the derivation above, we get the total similarity score:

$$
S = \frac{N}{2} \cdot \sin(\phi) - \frac{N}{2} \cdot \sin(\phi) = 0 \qquad \text{if } m \in \{0, N/2\}. \tag{5.10}
$$

Combining terms

Stepping back, if we take $m \notin \{0, N/2\}$, pick an arbitrary phase ϕ, generate a signal $x[n]$ according to Equation (5.9), and compare it to cosine and sine waveforms as above, we will observe the following similarity scores:

- $S(x, \cos) = \frac{N}{2} \cdot \cos(\phi)$

- $S(x, \sin) = -\frac{N}{2} \cdot \sin(\phi)$.

If we additionally scale x by an amplitude A, this scaling factor would carry through all of the summations above:

- $S(A \cdot x, \cos) = \frac{N}{2} \cdot A \cdot \cos(\phi)$

- $S(A \cdot x, \sin) = -\frac{N}{2} \cdot A \cdot \sin(\phi)$.

This is starting to look pretty interesting. However, it would be better if both scores had the same sign. We can accomplish this by changing which signals we compare to: instead of comparing to $\sin\left(2\pi \cdot \frac{m}{N} \cdot n\right)$, we'll compare to $-\sin\left(2\pi \cdot \frac{m}{N} \cdot n\right)$. Since this is just a multiplication by -1, like the amplitude scaling above, we can push it through the similarity calculation and obtain:

- $S(A \cdot x, \cos) = \frac{N}{2} \cdot A \cdot \cos(\phi)$

- $S(A \cdot x, -\sin) = \frac{N}{2} \cdot A \cdot \sin(\phi)$.

Now this is interesting.

We started with a sinusoid x having amplitude A and phase offset ϕ. The resulting scores can be interpreted as horizontal and vertical coordinates of a point with radius $A \cdot N/2$ and angle ϕ.

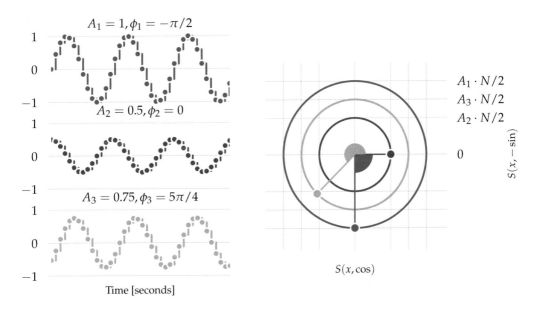

Figure 5.14 *Left*: three different waves of the form $x[n] = A \cdot \cos(2\pi \cdot n \cdot m/N + \phi)$ are compared to cos and $-\sin$ reference signals with the same frequency. *Right*: interpreting the similarity scores for each signal as horizontal and vertical coordinates places each one on a circle of radius $A \cdot N/2$ at angle ϕ.

Rather than carry around two coordinates for each frequency comparison, we can combine them into a single complex number. We will associate the comparison to cos with the **real** part, and the comparison to $-\sin$ with the **imaginary** part.

5.5.2 The discrete Fourier transform

Finally, we can formally define the discrete Fourier transform (DFT). The DFT takes as input a signal $x[n]$ of N samples, and produces a sequence $X[m]$ of N complex numbers representing amplitude and phase for each analysis frequency.

Definition 5.2 (The discrete Fourier transform (DFT)). For an arbitrary input signal $x[n]$ of N samples, its discrete Fourier transform is a sequence of N complex numbers $X[m]$ ($m = 0, 1, \ldots, N-1$) defined as follows:

$$X[m] = \sum_{n=0}^{N-1} x[n] \cdot \left(\cos\left(2\pi \cdot \frac{m}{N} \cdot n \right) - \mathrm{j} \cdot \sin\left(2\pi \cdot \frac{m}{N} \cdot n \right) \right) \qquad (5.11)$$

or, equivalently using Euler's formula (4.1),

$$X[m] = \sum_{n=0}^{N-1} x[n] \cdot \exp\left(-\mathrm{j} \cdot 2\pi \cdot \frac{m}{N} \cdot n \right). \qquad (5.12)$$

DFT notation

The $X[m]$ (capital) notation for the DFT of x is common, but not universal. Some other notations that you might see in the wild include

- $\hat{x}[m]$, and

- $\mathcal{F}(x)[m]$.

The DFT equation (5.11) can be implemented in code as follows:

```python
import numpy as np

def dft(x):
    '''The Discrete Fourier Transform

    Parameters
    ----------
    x : np.ndarray, The input signal

    Returns
    -------
    X : np.ndarray, same shape as x, The DFT sequence: X[m]
    '''

    N = len(x)  # Number of samples = number of frequencies
    X = np.zeros(N, dtype=np.complex)  # Allocate the output array

    for m in range(N):  # For each analysis frequency
        for n in range(N):  # and each sample
            # Compare to cos and -sin at this frequency
            X[m] = X[m] + x[n] * (np.cos(2 * np.pi * m / N * n)
                                  - 1j * np.sin(2 * np.pi * m / N
                                  * n))
    return X
```

Note that this code will be **extremely slow**, and should not be used for long signals. Instead, you should probably use a *fast Fourier transform* (FFT) implementation, like the one provided by `numpy`:

```
import numpy as np

X = np.fft.fft(x)
```

Some facts about the DFT

In the coming chapters, we will devote a significant amount of time to understanding everything we can about the DFT. For the time being, here is a brief list of basic facts that can be derived by a close examination of our process in building up the DFT.

For convenience, we will write $\mathrm{DFT}(x)$ to denote the entire sequence of DFT *components* $X[0], X[1], X[2], \ldots, X[N-1]$.

1. The DFT sequence $X[0], X[1], \ldots X[N-1]$ does not index *time* like the sample array $x[n]$ does. Instead, each $X[m]$ measures the contribution of a sinusoid at a particular frequency to the **entire signal** x.

2. The DFT sequence always has the same number of analysis frequencies as the number of samples N. The first $N/2$ correspond to positive frequencies, the last $N/2$ correspond to negative frequencies.

3. The $m = 0$ component is always a real number–this follows immediately from (5.10). It is known as the *direct current* (DC) component, and always equals the sum of the sample values.

4. For each $X[m]$, the magnitude $|X[m]|$ (`np.abs(X[m])`) corresponds to the amplitude of a sinusoid at frequency $f_s \cdot m/N$ present in the signal. Similarly, its angle in the complex plan, $\angle X[m]$ (`np.angle(X[m])`), corresponds to the **phase** of the sinusoid.

5. The sequence of values $|X[m]|$ is known as the **magnitude spectrum** of x.

6. The magnitude $|X[m]|$ is not the same as the real part $\mathrm{Re}X[m]$. Remember that for a complex number z,

$$|z| = \sqrt{(\mathrm{Re}z)^2 + (\mathrm{Im}z)^2}$$

Both the real and imaginary parts of the number contribute to both magnitude and phase.

5.5.3 Summary

Developing the DFT takes a lot of work. So does understanding it completely. **Don't panic if this section was difficult to get through.**

To finish off this chapter, the next section will show several examples signals and their corresponding DFTs.

5.6 EXAMPLES

In this section, we'll go in-depth with a few concrete example signals, and investigate their corresponding discrete Fourier transforms.

5.6.1 Impulse and delay

As a first example, let's look at an impulse:

$$x_|[n] = \begin{cases} 1 & \text{if } n = 0 \\ 0 & \text{otherwise.} \end{cases}$$

In this example, the DFT equation (5.12) gives us

$$
\begin{aligned}
X[m] &= \sum_{n=0}^{N-1} x_|[n] \cdot \exp\left(-\mathrm{j} \cdot 2\pi \cdot m \cdot n/N\right) \\
&= x_|[0] \cdot \exp\left(-\mathrm{j} \cdot 2\pi \cdot m \cdot 0/N\right) && \text{because } x_|[n \neq 0] = 0 \\
&= 1 && \text{because } \exp(0) = 1.
\end{aligned}
$$

So each DFT component has value $X[m] = 1 = 1 + 0\mathrm{j}$.

More generally, we can consider a d-step delay signal:

$$x[n] = \begin{cases} 1 & \text{if } n = d \\ 0 & \text{otherwise.} \end{cases}$$

which gives us a DFT sequence:

$$
\begin{aligned}
X[m] &= \sum_{n=0}^{N-1} x[n] \cdot \exp\left(-\mathrm{j} \cdot 2\pi \cdot m \cdot n/N\right) \\
&= x[d] \cdot \exp\left(-\mathrm{j} \cdot 2\pi \cdot m \cdot d/N\right) && \text{because } x[n \neq d] = 0 \\
&= \exp\left(-\mathrm{j} \cdot 2\pi \cdot m \cdot d/N\right).
\end{aligned}
$$

Fig. 5.15 illustrates this for a few different values of d.

Viewed as a sequence (over m), the DFT of a delayed impulse produces two sinusoids (one real, one imaginary) which cycle d times over the duration of the signal.

This will be useful later on when we revisit convolution!

5.6.2 Sinusoids

Since the DFT turns impulses into sinusoids, the next logical question to ask is what the DFT does to sinusoids? Luckily, we've already gone through the work of figuring this out in previous sections.

Let $x[n]$ be a sinusoid at analysis frequency index $k \notin \{0, N/2\}$ with phase ϕ and amplitude A:

$$x[n] = A \cdot \cos\left(2\pi \cdot \frac{k}{N} \cdot n + \phi\right)$$

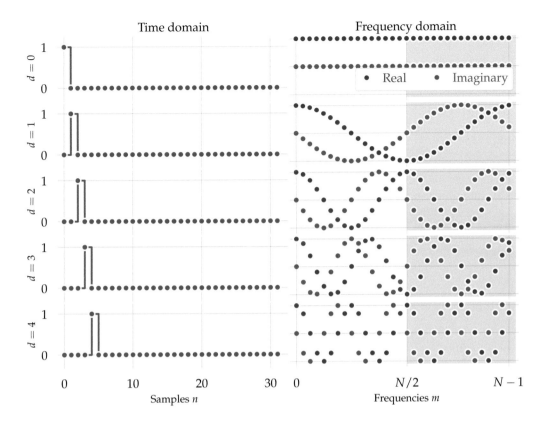

Figure 5.15 Each row shows the time-domain (left) and frequency-domain (right) representation of a d-step delay signal (for varying values of d. The shaded region corresponds to negative analysis frequencies.

For $m = k$, we will have

$$
\begin{aligned}
X[k] &= \frac{N}{2} \cdot A \cdot \left(\cos \left(\phi \right) + \mathrm{j} \cdot \sin(\phi) \right) \\
&= \frac{N}{2} \cdot A \cdot \exp \left(\mathrm{j}\phi \right). \qquad \text{By Euler's formula}
\end{aligned}
$$

But what about the rest of the spectrum?

First, let's look at $m = N - k$. For the real part of the reference signal, we'll have

$$
\begin{aligned}
\cos \left(2\pi \cdot \frac{N - k}{N} \cdot n \right) &= \cos \left(2\pi \cdot \frac{k - N}{N} \cdot n \right) && \cos(\theta) = \cos(-\theta) \\
&= \cos \left(2\pi \cdot \frac{k - N}{N} \cdot n + 2\pi \cdot \frac{N}{N} \cdot n \right) && \text{Add a whole number of rotations} \\
&= \cos \left(2\pi \cdot \frac{k - N + N}{N} \cdot n \right) && \text{Collect like terms} \\
&= \cos \left(2\pi \cdot \frac{k}{N} \cdot n \right) && \text{Cancel } N - N = 0,
\end{aligned}
$$

so the real part of $X[N - k]$ is the same as for $X[k]$. For the imaginary part, we'll get a sign flip:

$$-\sin\left(2\pi \cdot \frac{N - k}{N} \cdot n\right) = +\sin\left(2\pi \cdot \frac{k - N}{N} \cdot n\right) \qquad -\sin(\theta) = \sin(-\theta)$$

$$= \sin\left(2\pi \cdot \frac{k - N + N}{N} \cdot n\right)$$

$$= \sin\left(2\pi \cdot \frac{k}{N} \cdot n\right),$$

so the imaginary part of $X[N - k]$ is the *opposite* of $X[k]$. This gives us

$$X[N - k] = \frac{N}{2} \cdot A \cdot (\cos(\phi) - \mathrm{j} \cdot \sin(\phi))$$

$$= \frac{N}{2} \cdot A \cdot \exp(-\mathrm{j}\phi). \qquad \text{By Euler's formula}$$

For all other $m \neq k$, we will get $X[m] = 0$ as shown in Figure 5.16.

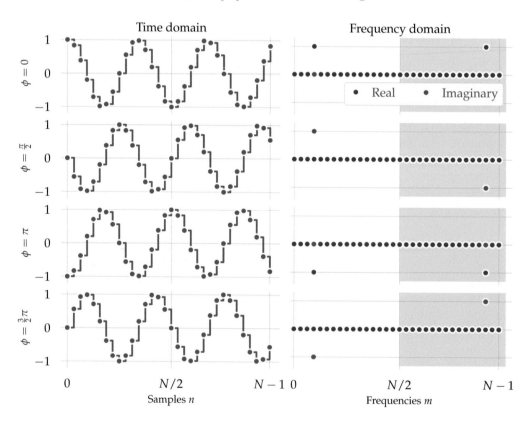

Figure 5.16 **Left**: sinusoids at analysis frequency $k = 3$ with varying phase ϕ. **Right**: the DFT of a sinusoid is a pair of impulses located at k and $N - k$. The height of the impulses depends on the phase ϕ.

5.6.3 A real sound

The examples above are helpful for building intuition about how the DFT behaves on idealized, synthetic signals. However, it's also helpful to see how the DFT behaves on a real signal. Here, we'll use the DFT to visualize a recording of a single note played on a trumpet.

```python
# We'll use soundfile to load the signal
import soundfile as sf

# And IPython to play it back in the browser
from IPython.display import Audio, display

# A single note (D#5) played on a trumpet
# https://freesound.org/s/48224/ - License: CC BY-NC 3.0
x, fs = sf.read('48224__slothrop__trumpetf3.wav')

N = len(x)   # How many samples do we have?

X = np.fft.fft(x)   # Compute its DFT

# and plot both the signal and its spectrum
fig, (ax_time, ax_freq) = plt.subplots(nrows=2)

time = np.arange(N) / fs   # get the sample times for x

ax_time.plot(time, x)   # plot the signal
ax_time.set(xlabel='Time [seconds]', title='Signal $x[n]$')

ax_freq.plot(np.abs(X), color=colors[11])   # plot the magnitudes␣
↪|X[m]|

# We'll shade the negative frequency range
ax_freq.axvspan(N/2, N, color='k', alpha=0.1, zorder=-1)
ax_freq.set(xlabel='Frequency index $m$', title='DFT magnitude
↪$|X[m]|$');
```

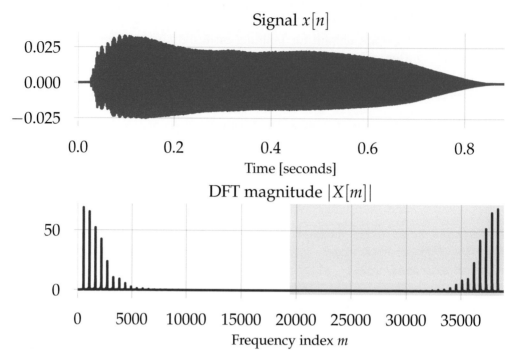

The first plot illustrates the raw waveform $x[n]$. While it's possible to discern some properties of the audio from the plot (e.g., how its amplitude rises and decays over time), it's virtually impossible to infer anything about the pitch or frequency content of the note.

The magnitude spectrum (second plot) shows some interesting structure, but it would be easier to understand if we use the actual analysis frequency values f_m rather than their index m for the horizontal axis.

We could calculate this manually using the rule

$$f_m = \begin{cases} \frac{m}{N} \cdot f_s & \text{if } 0 \leq m < N/2 \\ \frac{m-N}{N} \cdot f_s & \text{if } N/2 \leq m < N. \end{cases}$$

Fortunately, `numpy` gives a function that does exactly this for us: `np.fft.fftfreq`. It takes as input the number of samples N and the sample period $t_s = 1/f_s$. So we can instead do:

```
freqs = np.fft.fftfreq(N, 1/fs)
ax_freq.plot(freqs, np.abs(X))
```

From Fig. 5.17 we can observe that the magnitude spectrum $|X[m]|$ is sparse: it has small magnitudes for most frequencies (near 0). The first prominent peak (starting from 0 and looking to the right) that we can see is close to the frequency of the note: $D\sharp5$. If we want to find the actual frequency of the peak in the spectrum we can do so as follows:

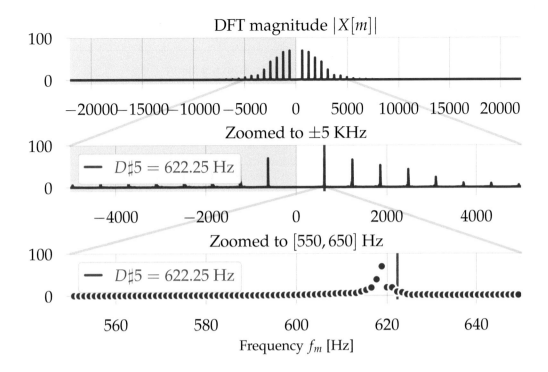

Figure 5.17 The DFT spectrum $|X[m]|$ of a trumpet playing D#5 = 622.25 Hz, plotted as a function of analysis frequency f_m. The middle and bottom plots contain the same information as the top, but zoomed into ±5 KHz (middle) and [550, 650] Hz (bottom). A prominent peak can be observed near $f_m \approx 622.25$ (and -622.25).

```
# get our DFT frequencies
frequencies = np.fft.fftfreq(N, 1./fs)

# find m index of largest magnitude
peak_m = np.argmax(np.abs(X))

# Get the f0
f0 = frequencies[peak_m]
```

Note that the peak is not *exactly* at the ideal pitch for D♯5 = 622.25: it lands about 10 cents flat at 618.8 Hz, with energy spread out around neighboring frequencies. This is still much closer to D♯5 than D5 = 587.33 though, so we would still perceive it as a D♯. This spread of energy is quite common, though it can be counter-intuitive the first time you see it. It happens because naturally generated frequencies are almost never exactly at our analysis frequencies. We'll come back to this point in the next chapter.

If we continue looking at increasing frequencies f_m (middle plot), we will find other peaks evenly spaced. These correspond to the *harmonics* (also known as *overtones* or *partials*) of the note being played.

Definition 5.3 (Harmonic series). For a given fundamental frequency f_0, its **harmonic series** is given by

$$f_k = (k+1) \cdot f_0 \qquad k = 1, 2, 3, \ldots$$

In the case of $f_0 = 622.25$, its harmonic series is the sequence $622.25, 1244.5,$ $1866.75, 2489.0, \ldots$. The amount of energy at each harmonic of a fundamental frequency differs from one instrument to the next, and this is part of what gives each instrument its distinctive *timbre*.

If the signal is relatively simple (e.g., an isolated source playing a single note), the DFT is a great way to quickly get a sense of its frequency content. Do not be deceived by this example though: most signals are *not* simple, and a visual inspection of the DFT is not generally sufficient to understand all signals.

5.6.4 Summary

We've now seen the DFT applied to several kinds of signals, producing different behaviors. Sometimes we can reason about the DFT analytically, as in the first two examples. Impulses and delays look like sinusoids in the frequency domain. Sinusoids look like (pairs of) impulses.

Other times, the best we can do is approach it empirically (as in the case of the trumpet recording), and observe the output of the DFT. With a little practice, one can learn to use the DFT to quickly get a sense of the contents of a signal that are not readily apparent from its waveform.

In the next few chapters, we'll dig deeper into the theoretical properties of the DFT.

5.7 SUMMING SINUSOIDS

In the previous sections, we've made use of the fact that the samples of a wave at an analysis frequency must sum to zero. However, we never actually proved this from first principles.

In fact, we can say something a little stronger, and exactly characterize the sum of samples for a wave at any frequency:

$$\sum_{n=0}^{N-1} \cos\left(2\pi \cdot f \cdot \frac{n}{f_s} + \phi\right) \quad = \quad ?$$

Computing this sum requires adding up wave samples, which can be done directly, though it can be tedious. As we'll see, this is a case where the complex exponential form is more convenient to work with.

5.7.1 Aside: geometric series

Recall that Euler's formula (4.1) converts between rectangular and polar coordinates:

$$e^{j \cdot \theta} = \cos(\theta) + j \cdot \sin(\theta).$$

Since this holds for any angle θ, it must also hold for $n \cdot \theta$ as we vary n:

$$e^{\mathrm{j} \cdot n \cdot \theta} = \cos(n \cdot \theta) + \mathrm{j} \cdot \sin(n \cdot \theta),$$

and by the product rule for exponents, we can re-write the left-hand side as follows:

$$e^{\mathrm{j} \cdot n \cdot \theta} = \left(e^{\mathrm{j} \cdot \theta}\right)^n.$$

This will allow us to turn a summation of wave samples into a summation of the form z^n, which is also known as a *geometric series*. Geometric series have many nice properties, but for now, the one we'll need is a formula for calculating the sum of the first N terms.

Lemma 5.1 (Summing finite geometric series). Let z be a complex number except 0 or 1, and let $N > 0$ be an integer. Then

$$\sum_{n=0}^{N-1} z^n = \frac{1 - z^N}{1 - z}.$$

In plain language, this lemma says that summing up increasing powers of a number z (even a complex number z) can be equivalently expressed as a ratio, rather than adding up individual terms.

Proof. To prove *Lemma 5.1*, observe that if we multiply the left-hand side (the summation) by $(1 - z)$, then successive terms in the summation will partially cancel each-other out.

$$
\begin{aligned}
(1 - z) \cdot \sum_{n=0}^{N-1} z^n &= \sum_{n=0}^{N-1} (1 - z) \cdot z^n && \text{Distribute } (1 - z) \\
&= \sum_{n=0}^{N-1} \left(z^n - z^{n+1}\right) && \text{Multiply by } z^n \\
&= \left(z^0 - z^1\right) + \left(z^1 - z^2\right) + \cdots + \left(z^{N-1} - z^N\right) && \text{Expand sum} \\
&= z^0 + \left(-z^1 + z^1\right) + \cdots + \left(-z^{N-1} + z^{N-1}\right) - z^N && \text{Re-group terms} \\
&= z^0 - z^N && \text{Cancel } \left(-z^k + z^k\right) \\
&= 1 - z^N && z \neq 0 \Rightarrow z^0 = 1
\end{aligned}
$$

Since $z \neq 1$ (by hypothesis), we know that $1 - z \neq 0$, so we can safely divide both sides of this equation by $1 - z$ to get the identity:

$$\sum_{n=0}^{N-1} z^n = \frac{1 - z^N}{1 - z}.$$

$\square$

Lemma 5.1 allows us to efficiently compute a sum $z^0 + z^1 + \ldots z^{N-1}$. In code, we could compute this iteratively, one value at a time:

```
S = 0
for n in range(N):
    S = S + z**n
```

or we can use the lemma to compute it one step:

```
S = (1 - z**N) / (1 - z)
```

The latter will be much faster, and the time it takes to compute the result will not depend on N.

5.7.2 Summing complex exponentials

Now that we have *Lemma 5.1*, we can state the following theorem.

Theorem 5.1 (Complex exponential sums). Let $\theta \neq 0$ be an angle, and let $N > 0$ be an integer. Then,

$$\sum_{n=0}^{N-1} e^{j \cdot n \cdot \theta} = 0 \quad \text{if and only if} \quad \theta \equiv 2\pi \cdot \frac{k}{N}$$

for an integer $k \neq 0 \mod N$.

> Note that we're using equivalence ($\equiv$) and not strict equality ($=$) here to compare angles θ. $\theta \equiv 0$ covers the cases $\theta \in \{0, 2\pi, -2\pi, 4\pi, -4\pi, \dots\}$.

In plain language, *Theorem 5.1* says that a complex sinusoid with a frequency that completes a whole number of cycles in N samples must sum to 0. An example of this is illustrated below in Fig. 5.18. We can actually say a bit more than that, and also characterize what happens for any frequency, and see that if the wave does not complete a whole number of cycles, then its sum cannot be 0.

Note that this does not handle the special case of $\theta \equiv 0$ ($f_0 = 0$) case, for which the summation simplifies to $1 + 1 + 1 + \dots = N + 0j$.

Proof

Theorem 5.1 has an "if and only if" form, so we'll need to prove both directions:

- ($\Rightarrow$) θ of the given form implies the summation must be zero, and

- ($\Leftarrow$) the summation being zero implies θ takes the given form.

Note that $e^{j \cdot n \cdot \theta} = \left(e^{j \cdot \theta}\right)^n$ by the product rule for exponents. If $\theta \neq 0$ then $e^{j \cdot \theta} \neq 1$, so we are allowed to use the geometric series identity with $z = e^{j \cdot \theta}$:

$$\sum_{n=0}^{N-1} e^{j \cdot n \cdot \theta} = \sum_{n=0}^{N-1} \left(e^{j \cdot \theta}\right)^n = \frac{1 - e^{j \cdot N \cdot \theta}}{1 - e^{j \cdot \theta}}. \tag{5.13}$$

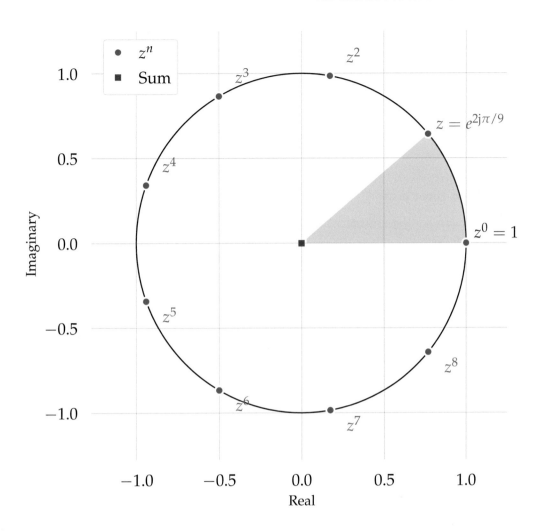

Figure 5.18 An example of the summation $\sum\limits_{n=0}^{N-1} z^n$ where $N = 9$ and $z = e^{2\pi \cdot j/9}$. The angle of z (shaded region) is $2\pi \cdot 1/N$, so this summation will total to 0.

Proof. ($\Rightarrow$)

If $\theta = 2\pi \cdot k/N$ for integer k, then the numerator can be equivalently expressed as

$$
\begin{aligned}
1 - e^{j \cdot N \cdot \theta} &= 1 - e^{j \cdot N \cdot 2\pi \cdot k/N} && \text{substitute } \theta \\
&= 1 - e^{j \cdot 2\pi \cdot k} && \text{cancel } N/N \\
&= 1 - e^{j \cdot 0} && \text{cancel extra rotations } 2\pi \cdot k \\
&= 1 - 1 \\
&= 0.
\end{aligned}
$$

Since the numerator is 0, so too is the entire summation. $\qquad\square$

Proof. ($\Leftarrow$)

In the other direction, if we assume the summation is 0, then we must have

$$0 = 1 - e^{j \cdot N \cdot \theta} \quad \Rightarrow \quad e^{j \cdot N \cdot \theta} = 1 = e^{j \cdot 0},$$

which implies $N \cdot \theta = 2\pi \cdot k$ for some integer k because $N \cdot \theta$ must be equivalent to a whole number of rotations (in either direction). Dividing through by N, we get $\theta = 2\pi \cdot k/N$. $\square$

5.7.3 What about phase?

We can generalize the statement above to handle phase offsets as well. Because the phase offset does not change with the sample index n, we can factor it out of the summation:

$$
\begin{aligned}
\sum_{n=0}^{N-1} e^{j \cdot (\theta \cdot n + \phi)} &= \sum_{n=0}^{N-1} e^{j \cdot \theta \cdot n} \cdot e^{j \cdot \phi} \\
&= e^{j \cdot \phi} \cdot \sum_{n=0}^{N-1} e^{j \cdot \theta \cdot n} \\
&= e^{j \cdot \phi} \cdot \frac{1 - e^{j \cdot N \cdot \theta}}{1 - e^{j \cdot \theta}}.
\end{aligned}
\tag{5.14}
$$

This says that when a wave is shifted by ϕ, the summation is multiplied by $e^{j \cdot \phi}$. Note that this is a pure rotation, so if the summation was 0 (i.e., we had an analysis frequency), then it will still be zero under *any* phase shift. Likewise, if the summation was non-zero, it will remain non-zero under any shift.

5.7.4 Back to the original question

The theorem above is for complex exponentials, which involve both a real and imaginary component. However, our original question was about summations of general waves in standard form:

$$\sum_{n=0}^{N-1} \cos\left(2\pi \cdot f \cdot \frac{n}{f_s} + \phi\right)$$

For this, we can use the fact that $\cos(\theta)$ is equivalent to the real part of $e^{j \cdot \theta}$, which implies

---

This is because for any complex number $z = a + j \cdot b$, we have

$$
\begin{aligned}
z + \overline{z} &= a + j \cdot b + a - j \cdot b \\
&= 2a,
\end{aligned}
$$

so the real part a can be extracted by adding z to its conjugate $\overline{z}$ and dividing by 2.

$$\cos(\theta) = \frac{1}{2} \cdot \left(e^{j \cdot \theta} + e^{-j \cdot \theta} \right),$$

or when a sample index n and phase offset ϕ are introduced, and letting $\theta = 2\pi \cdot f/f_s$:

$$\cos(\theta \cdot n + \phi) = \frac{1}{2} \cdot \left(e^{j \cdot (\theta \cdot n + \phi)} + e^{-j \cdot (\theta \cdot n + \phi)} \right).$$

By using (5.14), we can transform each part of this summation independently. The end result is the rather unwieldy formula:

$$\sum_{n=0}^{N-1} \cos(\theta \cdot n + \phi) = \frac{e^{j \cdot \phi}}{2} \cdot \frac{1 - e^{j \cdot N \cdot \theta}}{1 - e^{j \cdot \theta}} + \frac{e^{-j \cdot \phi}}{2} \cdot \frac{1 - e^{-j \cdot N \cdot \theta}}{1 - e^{-j \cdot \theta}} \qquad (5.15)$$

5.7.5 Why does this matter?

Many of the things we'd like to say about Fourier series depend on having waves "average out to zero". For continuous (time) signals, we can show this kind of thing via symmetry arguments (like in *chapter 1*), but when using discretely sampled signals, a bit more care must be taken.

The main theorem in this section tells us that waves at analysis frequencies always sum to 0, but in proving that theorem, we got as a byproduct a general equation (5.15) for sums of waves at arbitrary (non-analysis) frequencies and phase offsets. While this equation could be used in principle to bypass computing sums sample-by-sample, it is more useful as an analytical tool: it allows us to reason about the properties of the sum (e.g., whether it is 0 or non-zero, real or complex, etc.) just by knowing the wave's parameters.

5.8 EXERCISES

Exercise 5.1. If $N = 100$ and $f_s = 8000$, what is the lowest non-zero analysis frequency?

Exercise 5.2. What values of f_s and N would be necessary to ensure that the minimum and maximum (non-zero) analysis frequencies cover the range of typical human hearing (20-20000 Hz)?

Exercise 5.3. Imagine a signal of N samples is extended by padding with N additional zeros, resulting in a signal $y[n]$ of length $2 \cdot N$. How does the DFT $Y[m]$ compare to the DFT of the original signal $X[m]$?

In particular:

• How does the set of analysis frequencies change?

• For any given frequency index m, what can you say about $Y[m]$ in relation to the original spectrum X?

Exercise 5.4. Generate different types of signals with the same fundamental frequency (square, triangle, sawtooth). For each one, compute its DFT and look at its magnitude spectrum. Can you tell apart the different types of waves by the placement and magnitudes of their harmonics?

Hint. The `scipy.signal.square` and `scipy.signal.sawtooth` functions can be used to generate these signals.

Exercise 5.5. Generate a sinusoid at $f_0 = 110$ for 1 second at $f_s = 8000$. Then, apply different levels of *quantization*: 8-bit, 4-bit, 2-bit, and 1-bit. For each quantized signal, compute its DFT. How does it compare (qualitatively) to the original signal's DFT?

Exercise 5.6. Make (or find) a recording of the note $D\sharp 5$ on an instrument of your choice. Using the code provided for analyzing the *trumpet recording*, compute its DFT and magnitude spectrum. How does its spectrum compare to that of the trumpet?

## CHAPTER 6

Properties of the DFT

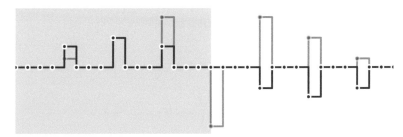

In this chapter, we'll go more in depth with the DFT to understand its fundamental properties. Specifically, we'll cover the following four topics:

- Linearity: what happens when we mix signals?

- The DFT shifting theorem: what happens when we delay signals?

- Conjugate symmetry: what is going on with negative frequencies?

- Spectral leakage: what happens to non-analysis frequencies?

A fair word of warning: this chapter will be technical and possibly a bit dry. However, a good understanding of these properties will help us reason about the DFT applied to more complicated signals, and eventually, what the DFT can tell us about filtering.

6.1 LINEARITY

The first major property of the DFT that we'll cover is **linearity**. We've already seen linearity in the context of *LSI systems*, where we used it to understand what happens when you convolve a filter h with a mixture of signals.

Here, we'll use linearity slightly differently, but the basic idea is the same.

Property 6.1 (DFT Linearity). For any pair of signals $x_1[n]$ and $x_2[n]$ of the same length, and real numbers c_1, c_2, the following holds:

$$\mathrm{DFT}(c_1 \cdot x_1 + c_2 \cdot x_2) = c_1 \cdot \mathrm{DFT}(x_1) + c_2 \cdot \mathrm{DFT}(x_2).$$

DOI: 10.1201/9781003264859-6

We'll prove this property algebraically, starting from the definition of the DFT (5.12):

$$X[m] = \sum_{n=0}^{N-1} x[n] \cdot \exp\left(-j \cdot 2\pi \cdot \frac{m}{N} \cdot n\right)$$

Proof. Let $0 \leq m < N$ be any frequency index. Using $x = c_1 \cdot x_1 + c_2 \cdot x_2$ as our input signal, its mth DFT component is

$$\sum_{n=0}^{N-1} (c_1 \cdot x_1[n] + c_2 \cdot x_2[n]) \cdot \exp\left(-j \cdot 2\pi \cdot \frac{m}{N} \cdot n\right)$$

$$= \sum_{n=0}^{N-1} \left(c_1 \cdot x_1[n] \cdot \exp\left(-j \cdot 2\pi \cdot \frac{m}{N} \cdot n\right)\right.$$

$$\left. + c_2 \cdot x_2[n] \cdot \exp\left(-j \cdot 2\pi \cdot \frac{m}{N} \cdot n\right)\right)$$

$$= c_1 \cdot \sum_{n=0}^{N-1} x_1[n] \cdot \exp\left(-j \cdot 2\pi \cdot \frac{m}{N} \cdot n\right)$$

$$+ c_2 \cdot \sum_{n=0}^{N-1} x_2[n] \cdot \exp\left(-j \cdot 2\pi \cdot \frac{m}{N} \cdot n\right)$$

$$= c_1 \cdot X_1[m] + c_2 \cdot X_2[m].$$

So if we scale and mix two input signals, the resulting DFT component is the same scaling and mixing of their individual DFT components $X_1[m]$ and $X_2[m]$. Since this holds for any m it must hold for all m, so the entire DFT sequence of the mixture is the mixture of the DFT sequences. This is what we set out to prove. □

What does linearity buy us?

DFT linearity is important because most interesting signals are *not* just simple sinusoids. What DFT linearity says is that if we can represent an arbitrary signal x as a weighted combination of sinusoids, then we can reason about its DFT in terms of its constituent sinusoids.

As we will see in the next chapter (and as we've hinted at earlier), it turns out that *all* signals can be represented as a weighted combination of sinusoids.

6.1.1 Magnitude is not linear

Note that DFT linearity applies to the **complex numbers** $X[m]$ which encode both magnitude and phase. There is nothing wrong with adding complex numbers, provided it is done correctly, keeping the real and imaginary parts separate.

However, it is a common mistake to add **magnitudes** rather than the full complex numbers, as shown in Figure 6.1.

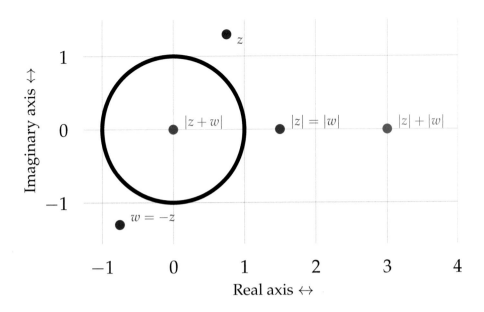

Figure 6.1 Complex numbers z and $w = -z$ have the same magnitude $|z| = |w|$ but different phase. The sum of their magnitudes $|z| + |w|$ is not the same as the magnitude of their sum $|z + w| = 0$.

Note (Magnitude is not linear.). If signals $x_1[n]$ and $x_2[n]$ have DFT series $X_1[m]$ and $X_2[m]$, then the sum $y[n] = x_1[n] + x_2[n]$ has DFT series $Y[m] = X_1[m] + X_2[m]$.

But remember,

$$|Y[m]| = |X_1[m] + X_2[m]|$$

$$\neq |X_1[m]| + |X_2[m]|.$$

Example 6.1. For an extreme case, consider $x_2[n] = -x_1[n]$ for some arbitrary, non-empty signal x_1. DFT linearity says that $X_2[m] = -1 \cdot X_1[m]$, so the spectra X_1 and X_2 are opposites. However, the spectral *magnitudes* are identical:

$$X_2[m] = -X_1[m] \quad \Rightarrow \quad |X_2[m]| = |-1 \cdot X_1[m]| = |X_1[m]|.$$

If we were to add the magnitude spectra, we'd get

$$|X_1[m]| + |X_2[m]| = |X_1[m]| + |X_1[m]| = 2 \cdot |X_1[m]|.$$

However, if we mix the two signals in the time domain, we get the empty signal:

$$y[n] = x_1[n] + x_2[n] = x_1[n] - x_1[n] = 0.$$

The empty signal has an empty spectrum: $Y[m] = 0$ for all m, but

$$|Y[m]| = 0 \neq 2 \cdot |X_1[m]|$$

in general.

What this says is that **mixing signals** does not equate to **mixing DFT magnitudes**.

6.2   THE DFT SHIFTING THEOREM

We've seen that for signals x which are pure sinusoids at an analysis frequency of index m, the corresponding DFT component $X[m]$ encodes the signal's amplitude and phase as a single complex number.

But what can we say about phase for more general signals? In particular, if we have an arbitrary signal x (not necessarily a sinusoid), what happens if we delay it?

Before we continue down this line of thought, we must first establish what it means to delay a signal under our periodicity assumption. This leads us to the notion of a **circular shift**.

Definition 6.1 (Circular shifting). Let $x[n]$ be a signal of N samples with DFT series $X[m]$, and define

$$y[n] = x[n - d \mod N]$$

to be the **circular shift** of x by d samples.

When it is clear from context that shifting is circular, we may drop the " $\mod N$" notation and simply write $y[n] = x[n - d]$.

Circular shifting, shown in Figure 6.2, can seem like a strange thing to do: taking samples from the end of the signal and putting them at the beginning? However, it is a natural consequence of combining our *periodicity assumption* with the definition of delay.

Indeed, it can lead to some strange behaviors if $x[n]$ is discontinuous at the repetition boundaries.

Given this definition of shifting (delay), we can say the following about its effect on the DFT spectrum of a signal.

Theorem 6.1 (DFT shifting). Let $x[n]$ be a signal with DFT spectrum $X[m]$, and let $y[n] = x[n - d]$ be the circular shift of x by d samples.

The DFT spectrum of $y[n]$ is given by:

$$Y[m] = X[m] \cdot \exp\left(-\mathrm{j} \cdot 2\pi \cdot \frac{m}{N} \cdot d\right).$$

This says that no matter what delay we use, it's always possible to exactly predict the spectrum of the delayed signal from the spectrum of the input signal.

The proof of the shifting theorem is purely algebraic, and relies on the periodicity assumption of x to allow a change of variables in the similarity calculation.

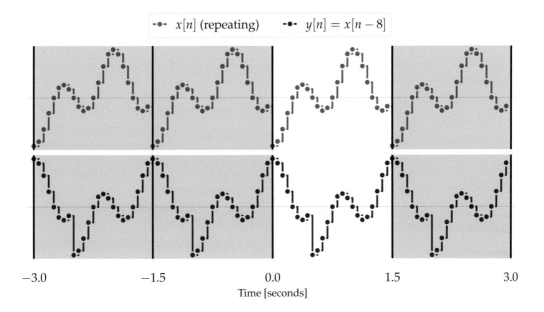

Figure 6.2 A repeating signal $x[n]$ with repetition boundaries (vertical lines) is shifted by 8 samples to produce $y[n]$; three additional repetitions are illustrated in the shaded regions. Because $x[n]$ is discontinuous at the repetition boundary, this discontinuity is shifted to the middle of $y[n]$, appearing as a large vertical gap in the signal.

Proof. If we have an index n, we will introduce a new variable $k = n - d$, and equivalently, $n = k + d$. In this case, $n = 0$ corresponds to $k = -d$, and $n = N - 1$ corresponds to $k = N - 1 - d$. Then we can calculate the DFT spectrum as follows

$$Y[m] = \sum_{n=0}^{N-1} y[n] \cdot \exp\left(-j \cdot 2\pi \cdot \frac{m}{N} \cdot n\right) \qquad \text{By definition of DFT}$$

$$= \sum_{n=0}^{N-1} x[n-d] \cdot \exp\left(-j \cdot 2\pi \cdot \frac{m}{N} \cdot n\right) \qquad y[n] = x[n-d]$$

$$= \sum_{k=-d}^{N-1-d} x[k] \cdot \exp\left(-j \cdot 2\pi \cdot \frac{m}{N} \cdot (k+d)\right) \qquad k = n-d$$

$$= \sum_{k=-d}^{N-1-d} x[k] \cdot \exp\left(-j \cdot 2\pi \cdot \frac{m}{N} \cdot k\right) \cdot \exp\left(-j \cdot 2\pi \cdot \frac{m}{N} \cdot d\right) \qquad e^{a+b} = e^a \cdot e^b$$

$$= \exp\left(-j \cdot 2\pi \cdot \frac{m}{N} \cdot d\right) \cdot \left(\sum_{k=-d}^{N-1-d} x[k] \cdot \exp\left(-j \cdot 2\pi \cdot \frac{m}{N} \cdot k\right)\right) \qquad \text{Delay part is constant}$$

$$= \exp\left(-j \cdot 2\pi \cdot \frac{m}{N} \cdot d\right) \cdot X[m] \qquad \text{definition of DFT.}$$

The last step follows because even though the summation ranges from $k = -d$ to $k = N - 1 - d$, it still computes the full DFT of x, just in a different order. □

6.2.1 What does this do?

It's worth taking some time to understand the shifting theorem. Not only does it say exactly how the spectrum changes, as a signal shifts in time, but the change itself is also interesting.

Note that the argument of the exponential is a purely imaginary number:

$$-\mathrm{j} \cdot 2\pi \cdot \frac{m}{N} \cdot d.$$

This means that multiplication implements a **rotation** of $X[m]$ in the complex plane. Its phase can change, but the magnitude must be the same.

Of course, this is exactly what we would hope should happen: delay only changes the horizontal (time) position of a signal, not its amplitude.

A slightly more subtle point is that each DFT component $X[m]$ changes **by a different amount**: the rotation depends on both m and d.

6.2.2 What does this not do?

The shifting theorem implies that if a signal is circularly shifted, its magnitude spectrum is unchanged:

$$y[n] = x[n - d] \qquad \Rightarrow \qquad |Y[m]| = |X[m]|.$$

However, the converse is **not true**: as shown in Figure 6.3, two signals with the same magnitude spectrum may not be related by a circular shift of one another.

Example 6.2. Let $x[n] = 1, 0, 0, 0, 0, \ldots$ denote an impulse with N samples, and let $y = -x$ denote its negation. As shown in the *previous chapter*, an impulse signal has spectrum $X[m] = 1$ for all m. By *DFT linearity*, we also know that $y = -x$ implies $Y[m] = -X[m]$. As a result, we have that

$$|Y[m]| = |-X[m]| = |X[m]| = 1 \quad \text{for } m = 0, 1, \ldots, N-1$$

so the two signals have the same magnitude spectrum. However, there is no delay d that turns x into y.

Tip. Phase is important. Having the same magnitude spectrum is not enough to ensure that two signals are the same except for a phase shift.

6.3 CONJUGATE SYMMETRY

Theorem 6.2 (DFT conjugate symmetry). Let $x[n]$ be a real-valued signal with N samples.

Then the DFT series $X[0], X[1], \ldots, X[N-1]$ has **conjugate symmetry**:

$$X[m] = \overline{X[N - m]}.$$

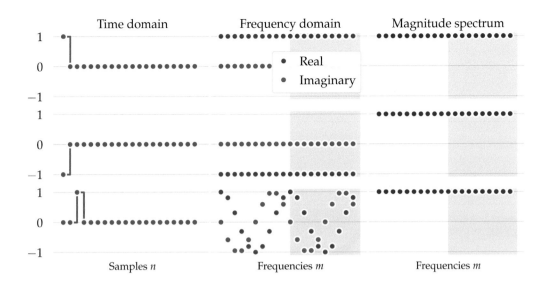

Figure 6.3 *Left*: an impulse (top), a negative impulse (middle), and a 3-step delay (bottom). *Center*: the DFT spectrum (real and imaginary) components of each signal are distinct. *Right*: all three signals have identical magnitude spectra $|X[m]| = 1$.

In words, this says that the DFT component $X[m]$ is the complex conjugate of component $X[N - m]$ (and vice versa): their real parts are identical, and their imaginary parts are negatives of each-other.

As it turns out, we've seen a version of this already when computing the DFT of *sinusoids*. A careful look at how we approached the problem will reveal that none of the argument depended on the fact that the input signal $x[n]$ was a sinusoid, only that it was real-valued. Here, we can give a slightly more streamlined proof using the complex exponential.

Proof. We'll start by observing that for any complex exponential $e^{\mathrm{j}\cdot\theta}$, the following holds:

$$\overline{e^{\mathrm{j}\theta}} = e^{-\mathrm{j}\theta}.$$

This is illustrated in Figure 6.4, and follows immediately from Euler's formula (4.1)

$$e^{\mathrm{j}\theta} = \cos(\theta) + \mathrm{j}\cdot\sin\theta$$

and the *symmetry rules* for cos and sin:

$$\cos(-\theta) = \cos(\theta)$$
$$\sin(-\theta) = -\sin(\theta)$$

so that

$$e^{-\mathrm{j}\cdot\theta} = \cos(-\theta) + \mathrm{j}\cdot\sin(-\theta)$$
$$= \cos(\theta) - \mathrm{j}\cdot\sin(\theta)$$
$$= \overline{e^{\mathrm{j}\cdot\theta}}.$$

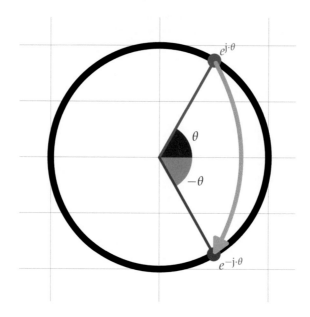

Figure 6.4 Conjugating a complex exponential $e^{j\theta}$ is equivalent to negating its angle θ.

Taking $\theta = 2\pi \cdot \frac{N-m}{N} \cdot n$, we get

$$\exp\left(-j \cdot 2\pi \cdot \frac{N-m}{N} \cdot n\right) = \exp\left(+j \cdot 2\pi \cdot \frac{m-N}{N} \cdot n\right) \qquad \text{cancelling negatives}$$

$$= \exp\left(j \cdot 2\pi \cdot \frac{m}{N} \cdot n\right) \qquad \text{add rotations } 2\pi \cdot n \cdot N/N$$

$$= \overline{\exp\left(-j \cdot 2\pi \cdot \frac{m}{N} \cdot n\right)} \qquad \text{conjugation.}$$

This lets us express $X[N - m]$ as follows:

$$X[N - m] = \sum_{n=0}^{N-1} x[n] \cdot \exp\left(-j \cdot 2\pi \cdot \frac{N-m}{N} \cdot n\right)$$

$$= \sum_{n=0}^{N-1} x[n] \cdot \overline{\exp\left(-j \cdot 2\pi \cdot \frac{m}{N} \cdot n\right)}$$

$$= \overline{\sum_{n=0}^{N-1} x[n] \cdot \exp\left(-j \cdot 2\pi \cdot \frac{m}{N} \cdot n\right)}$$

$$= \overline{X[m]}.$$

□

Warning. The last step hinges upon the fact that $x[n]$ is real-valued, so $x[n] \cdot \overline{z} = \overline{x[n] \cdot z}$ for any complex z. If $x[n]$ was complex-valued, this step would not go through, and we would not have conjugate symmetry.

6.3.1 Which frequencies do we need to compute?

DFT conjugate symmetry says that there is *redundancy* in the spectral content of a real-valued signal. But what exactly is the minimal set of frequencies that we *must* compute to fully represent the signal?

We can think of this by working through each $m = 0, 1, 2, \ldots N - 1$ and seeing which other components can be inferred once we have $X[m]$. We can stop when all N have been covered.

- $m = 0 \Leftrightarrow N - 0 = N \equiv 0$ - The DC component pairs with itself.

- $m = 1 \Leftrightarrow N - 1$

- $m = 2 \Leftrightarrow N - 2$

- $\ldots$

How far we can go with this depends on whether N is even or odd. If N is even, then

$$m = \frac{N}{2} \Leftrightarrow N - \frac{N}{2} = \frac{N}{2}$$

give us the Nyquist frequency, which like the $m = 0$ case, pairs with itself. In this case, we'll have a total of $M = 1 + N/2$ frequencies: $m = 0$ followed by everything up to and including $m = N/2$.

If N is odd, then there is no m such that $m = N - m$. The largest m such that $m < N/2$ (so $N - m > N/2$) is $m = (N - 1)/2$. This gives us a total of $M = 1 + (N - 1)/2$ frequencies.

These two cases can be combined to give a single formula for M

$$M = 1 + \left\lfloor \frac{N}{2} \right\rfloor$$

where $\lfloor \cdot \rfloor$ is the "floor" operation: the largest integer no greater than its input. This expression is implemented in Python as:

```
M = 1 + N // 2   # Integer division
```

Fig. 6.5 depicts the conjugate symmetry relationships graphically for both the even- and odd-length cases.

6.3.2 Why does this matter?

Conjugate symmetry can buy us some efficiency in both the time to compute the DFT, and the space it takes to store the results.

Most Fourier transform implementations provide a specialized function **rfft** (real-valued input FFT) which can efficiently compute just the non-negative frequency values.

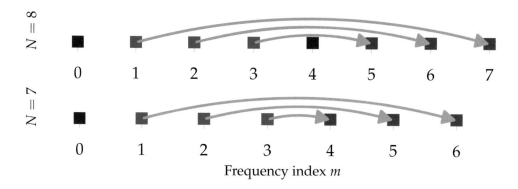

Figure 6.5 The DFT conjugate symmetry property relates $X[m]$ with $X[N-m]$ (arrows). Indices $m = 0$ and $m = N/2$ (if N is even) are only related to themselves. If the signal length is even (*top*, $N = 8$), then the first $M = 1 + N/2$ components are sufficient to determine all N components. If the signal length is odd (*bottom*, $N = 7$), then $M = 1 + (N-1)/2$ are sufficient.

```
# Compute the full spectrum, all N frequencies
X = np.fft.fft(x)


# Compute just the non-negative spectrum: 1 + N//2 frequencies
X = np.fft.rfft(x)
```

For small values of N like the examples we've been working with, this is not such a big deal. However, it is not uncommon to have signals of thousands of samples: in this case, reducing the amount of data by 50% can substantially cut down on memory consumption and processing time. This is also beneficial in real-time audio applications, where there can be tight constraints on latency and computing a DFT is often the computational bottleneck.

Additionally, conjugate symmetry teaches us two facts about the DFT:

1. Because the DC component ($m = 0$) is its own conjugate, $X[0] = \overline{X[0]}$, we know that it must be real-valued. This shouldn't be surprising, given that it must also equal the sum of the (real-valued) input samples contained in $x[n]$.

2. If N is even, then the Nyquist component $m = N/2$ is also self-conjugate: $X[N/2] = \overline{X[N/2]}$. It must also be real-valued if $x[n]$ is. This is much less obvious from the DFT definition.

6.4 SPECTRAL LEAKAGE AND WINDOWING

In this section, we'll take a more in-depth look at how the DFT responds to non-analysis frequencies, and how we might counteract these effects.

In an earlier section, we've seen that waves at non-analysis frequencies can produce sharp discontinuities at the boundaries of the signal (Fig. 6.2). If we were to listen to a signal like this on a repeating loop, these discontinuities would become audible as *clicks* or *pops*, not dissimilar to the sound of an impulse.

To illustrate the point, here we'll construct a pure sinusoid at $f = 111$, with $f_s = 1000$, and sample for 1/4 second ($N = f_s/4 = 250$). This produces the following wave:

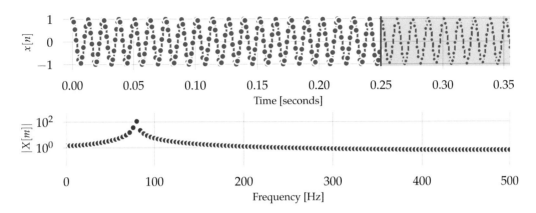

Figure 6.6 *Top*: A wave $x[n]$ at frequency $f = 79$, with sampling rate $f_s = 1000$ and duration of 1/4 second ($N = f_s/4$). $f = 79$ is not an integer multiple of $f_s/N = 4$, so it is not an analysis frequency. Note the large gap between samples when the signal repeats at $t = 0.25$. *Bottom*: The DFT magnitude spectrum $|X[m]|$, displayed on a logarithmic amplitude scale.

If you were to listen to this excerpt as a continuous loop, you would hear a tone with audible glitches every 1/4 second. The following code can be used to generate this behavior:

```
from IPython.display import Audio
import scipy.signal

fs = 1000
N = fs//4
f = 79

t = np.arange(N) / fs
x = np.tile(np.cos(2 * np.pi * f * t), 8)

y = scipy.signal.resample(x, 8 * len(x))
Audio(data=y, rate=8*fs)
```

To understand this behavior in the frequency domain, it can be helpful to think about the DFT of a unit impulse or delay signal (see Section 5.6.1). Impulses–and more generally, sharp discontinuities–are not easy to express with smoothly varying sinusoids, and this is why it takes the entire set of analysis frequencies to represent a delay signal (all $X[m]$ are non-zero in Fig. 5.15).

A similar thing happens to non-analysis frequencies: to explain the discontinuity at the boundary of the signal, the DFT uses the entire frequency spectrum, again producing non-zero DFT magnitudes $|X[m]|$ across the frequency spectrum. This phenomenon is called **spectral leakage**: the energy associated with our non-analysis frequency f has "leaked" over the entire spectrum.

The bad news is that spectral leakage cannot be avoided in general. The energy in a signal associated with each frequency has to go somewhere in the DFT, and if the frequency does not correspond to one of our analysis frequencies, then it will spread out.

The good news is that we can, to some extent, control leakage to direct the leaked energy in various ways. This is accomplished by a technique called **windowing**.

6.4.1 Windowing

The idea of windowing a signal is to force continuity at the boundaries, prior to performing the DFT. The simplest way to achieve this is by multiplying the signal $x[n]$ by another signal $w[n]$ of the same duration, such that $w[0] \approx w[N-1] \approx 0$, resulting in the *windowed DFT* $\hat{X}$:

$$\hat{X} \leftarrow \mathrm{DFT}(x[n] \cdot w[n])$$

Definition 6.2 (Hann window). A (discrete) *Hann window* of N samples is defined by the following equation:

$$w[n] = \sin^2\left(\frac{\pi n}{N}\right)$$

for $n = 0, 1, 2, \ldots, N-1$.

Hann windowing was developed by Julius Hann for smoothing climatological data [VHW03], and was later adapted by Blackman and Tukey for signal processing [BT58]. An example of a Hann window is given in Fig. 6.7.

Taking the DFT after windowing the signal significantly reduces the magnitude of components $\hat{X}[m]$ when m is far from the true frequency of the input signal, but retains energy for m close to the true frequency. For example, Fig. 6.8 illustrates the effect of applying a Hann window to a sinusoid.

If we loop the windowed signal, rather than the original signal, we'll see that the boundary discontinuities have vanished.

As shown in Fig. 6.9, the elimination of discontinuities does change the signal: we have also introduced a low-frequency amplitude modulation to the looped signal.

This effect is certainly audible. If we apply windowing to the previous example ($f = 111$ Hz sampled at $f_s = 1000$), and listen to the repeating signal as before, each loop of the signal now has a *pulse* due to the window.

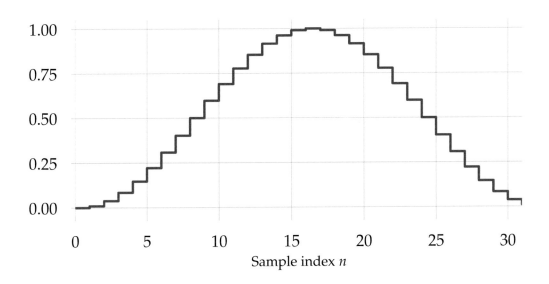

Figure 6.7 A Hann window of length $N = 32$.

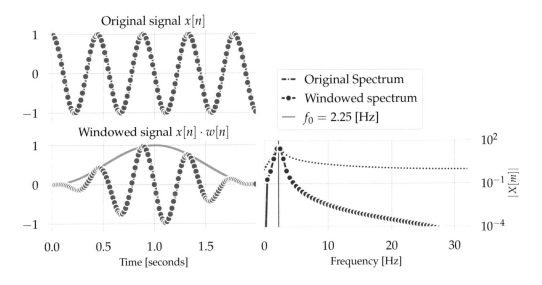

Figure 6.8 A sinusoid $x[n]$ at a non-analysis frequency (*upper-left*, $f = 2.25$, $N = 128$, $f_s = 64$) produces spectral energy at all frequencies in the DFT (*original spectrum*, bottom-right). Multiplying the sinusoid by a window function $w[n]$ (*bottom-left*) tapers the signal values to 0 at the beginning and end, and reduces spectral leakage in the DFT magnitude spectrum $|\hat{X}|$ (*windowed spectrum*, bottom-right).

```
from IPython.display import Audio
fs = 1000
N = fs//4
```

(continues on next page)

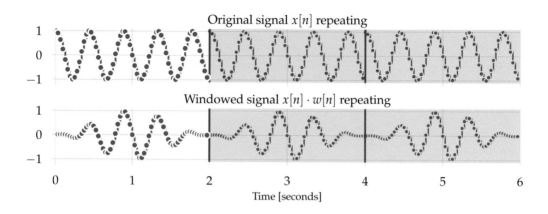

Figure 6.9 *Top*: repeating a sinusoid x at a non-analysis frequency produces discontinuities when the signal repeats ($t = 2, 4$). *Bottom*: windowing the sinusoid prior to repetition eliminates discontinuities but introduces low-frequency modulation.

(continued from previous page)

```
f = 79

t = np.arange(N) / fs
w = scipy.signal.get_window('hann', N, fftbins=False)
x = np.cos(2 * np.pi * f * t)
x = np.tile(x * w, 8)
y = scipy.signal.resample(x, 8 * len(x))
Audio(data=y, rate=8*fs)
```

In practice, one would not typically use windowing in this fashion, but it is helpful to listen to the looped signal to get a better intuition for what the signal looks like to the DFT. In this example, we have traded off transient discontinuities for a smoothly varying pulse, which can be more easily modeled by sinusoids.

6.4.2 Choosing a window

In our first example, we used a Hann window, which is essentially a carefully tuned cosine wave, but there are many, many, many other options. Most window functions have non-linear curves, and often end up resembling a "bell curve".

We won't go into the details of how each of these window functions are defined, but Fig. 6.10 demonstrates a handful of commonly used windows.

Fig. 6.10 demonstrates two key properties of windowing functions. First, different window functions will attenuate distant frequencies differently. The height of the spectrum in the last plot (Blackman-Harris) is around 10^{-4}, while the Hamming window is approximately 100x higher at 10^{-2}. From this, we might conclude that the Blackman-Harris window is "better" than the Hamming window, but we shouldn't be too hasty.

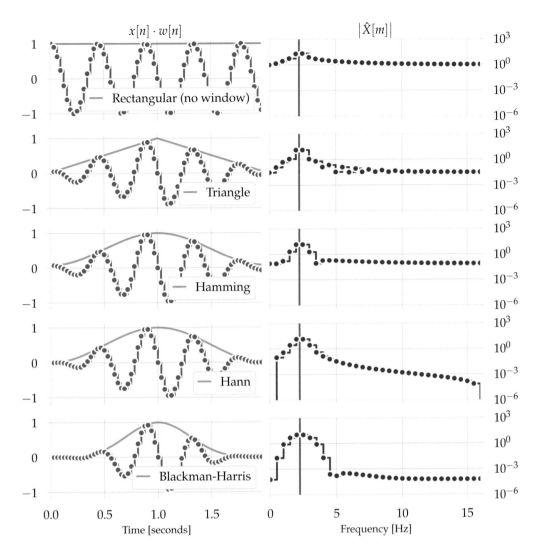

Figure 6.10 Different choices of window function w and the corresponding DFT magnitude spectrum $|\hat{X}|$ after applying each window to a sinusoid at non-analysis frequency $f = 2.25$ Hz.

The second property has to do with how much the energy spreads around its peak in the spectrum, the so-called "main lobe" width. From this perspective, the Blackman-Harris window has a broader main-lobe (around the peak frequency 2.25) than the other windows, so it might not be the best choice if our goal is to distinguish between nearby frequencies.

Tip. As a general rule, the Hann window is a good default choice for most audio applications.

What about analysis frequencies?

All of this was motivated by the problem of transients being induced by looping signals with non-analysis frequencies. In general, we won't know if a signal contains non-analysis frequencies, so it's natural to ask what would happen if we apply windowing in general? What happens to signals that actually do contain analysis frequencies?

Fig. 6.11 illustrates exactly the same comparisons as above, but now using an input signal x generated by a sinusoid at an analysis frequency.

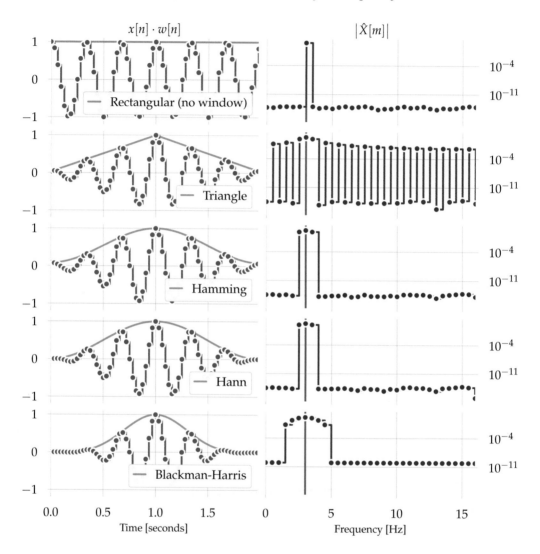

Figure 6.11 Different choices of window function w and the corresponding DFT magnitude spectrum after applying each window to a sinusoid at **analysis** frequency $f = 3$ Hz.

The spectral plots in Fig. 6.11 illustrate another price that we must pay to counteract leakage: if we do have analysis frequencies in the signal, applying windowing will spread some of their energy across the spectrum. Although in all cases, the energy far from the fundamental frequency ($f = 3$) is small (numerically close to 0), the energy in the windowed signals is dispersed around the peak, rather than being concentrated like in the un-windowed case.

One way to view this trade-off is that windowing reduces the distinction between analysis and non-analysis frequencies: both end up leaking across the spectrum, but the choice of window function allows us to control this behavior. In reality, almost no naturally occurring signals will line up precisely to the parameters of your signal analysis, so it's safer to assume that all energy is coming from non-analysis frequencies anyway.

6.4.3 Windowing in practice

Most signal processing frameworks provide a library of pre-defined window functions. In Python, these are provided by the function `scipy.signal.get_window`. To use a windowing function as we did in the example above, one first constructs the window of a given length, and then applies the DFT:

```
# We'll assume the input signal x already exists, and get its length
N = len(x)

# Build the window
w = scipy.signal.get_window('hann', N)

# Multiply by w and take the DFT
X = np.fft.rfft(x * w)
```

6.4.4 Summary

In this section, we've seen that windowing can help reduce the bad effects of spectral leakage. This doesn't necessarily mean that windowing should always be used, however, as it does alter the content of the signal.

As we will see later on, the main application of windowing has to do with the *short-time Fourier transform*, where a long signal is carved into small pieces for analysis purposes.

6.5 EXERCISES

Exercise 6.1. In general, magnitude is not a linear operation for complex numbers: $|w + z| \neq |w| + |z|$. Are there any special cases where this equality does hold? That is, can you find two complex numbers w and z such that $|w + z| = |w| + |z|$? If so, what general property must these numbers satisfy to ensure that equality holds?

Exercise 6.2. The DFT shifting theorem gives an equation to express $Y[m]$ as a complex rotation of $X[m]$ when $y[n]$ is a cyclic shift of $x[n]$. For a fixed frequency index m and signal length N, can you find a delay $d \notin \{0, N\}$ such that $X[m] = Y[m]$? That is, is there a non-trivial delay that leaves $X[m]$ unchanged?

Exercise 6.3. Let $y[n] = x[N - n]$ be the time-reversal of $x[n]$. What can you say about how the DFT $Y[m]$ relates to $X[m]$?

Exercise 6.4. Create (or find) a recording of an instrument playing a single, isolated note. Analyze the recording using:

- No windowing function: `np.fft.rfft(x)`

- Hann window: `np.fft.rfft(x * scipy.signal.hann(len(x), sym=False))`

- Two other window functions of your choice: e.g., Hamming, triangular, Kaiser, Blackman-Harris, etc.

How do the resulting spectra compare to each other, qualitatively? In particular, how do the height and widths of the spectral peaks change when different windows are used?

## CHAPTER 7

DFT invertibility

In this short chapter, we'll derive the *inverse* of the discrete Fourier transform.

As we've seen previously, the DFT maps a time-domain input $x[n]$ to a frequency domain output $X[m]$. The inverse DFT reverses this process, mapping the frequency domain input $X[m]$ to the time-domain output $x[n]$.

7.1   WARM-UP: A SINGLE SINUSOID

When we defined the discrete Fourier transform (DFT) (Section 5.5.2), we saw that it exactly encodes the amplitude (A) and phase (ϕ) of a sinusoidal input, at least when the frequency matches one of the analysis frequencies. Specifically, we saw that (for $m \notin \{0, N/2\}$),

$$x[n] = A \cdot \cos\left(2\pi \cdot \frac{m}{N} \cdot n + \phi\right) \quad \Rightarrow \quad X[m] = A \cdot \frac{N}{2} \cdot e^{j \cdot \phi}.$$

Of course, this doesn't tell us much if $x[n]$ is *not* of that particular given form. The DFT is still well defined, even if $x[n]$ is a completely arbitrary set of sample values:

$$X[m] = \sum_{n=0}^{N-1} x[n] \cdot e^{-2\pi \cdot j \cdot m \cdot n / N}$$

This raises the question: how do we interpret the coefficients $X[m]$ for arbitrary signals $x[n]$? We'll answer this by showing that there exists an *inverse* DFT, which can reconstruct the input signal $x[n]$ exactly from only its DFT coefficients $X[m]$.

7.1.1   Recovering a sinusoid

Before defining the full inverse DFT, let's first try to recover a single sinusoid from its DFT spectrum.

Recall from Section 5.6.2 that a sinusoid at an analysis frequency index m will have two non-zero DFT coefficients: $X[m]$ and $X[N - m]$, which are complex conjugates of each other: their real parts are the same, and their imaginary parts are

DOI: 10.1201/9781003264859-7

opposite:

$$X[m] = A \cdot \frac{N}{2} \cdot e^{\mathrm{j} \cdot \phi}$$

$$X[N - m] = A \cdot \frac{N}{2} \cdot e^{-\mathrm{j} \cdot \phi}$$

Each of these components can be used to produce a complex sinusoid $z_m[n]$ as follows:

$$z_m[n] = X[m] \cdot e^{2\pi \cdot \mathrm{j} \cdot \frac{m}{N} \cdot n}$$

$$= A \cdot \frac{N}{2} \cdot e^{\mathrm{j} \cdot \phi} \cdot e^{2\pi \cdot \mathrm{j} \cdot \frac{m}{N} \cdot n}$$

$$= A \cdot \frac{N}{2} \cdot e^{\mathrm{j} \cdot \left(2\pi \cdot \frac{m}{N} \cdot n + \phi\right)}$$

This is demonstrated visually in Fig. 7.1.

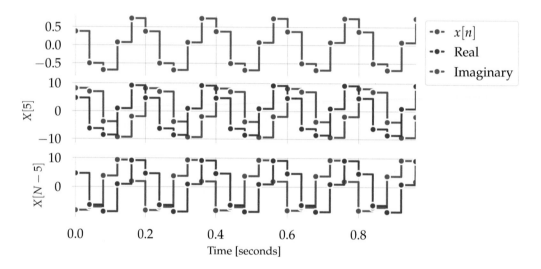

Figure 7.1 *Top*: a sinusoid $x[n] = 0.75 \cdot \cos(2\pi \cdot 5 \cdot \frac{n}{N} + \frac{\pi}{3})$ with $N = f_s = 25$. *Middle*: the complex sinusoid corresponding to $X[5]$ (real and imaginary portions shown separately). *Bottom*: the complex sinusoid corresponding to $X[N - 5]$. For complex sinusoids, the real and imaginary parts are plotted separately.

Because the imaginary parts of $X[m]$ and $X[N - m]$ are opposite ($X[m] = \overline{X[N - m]}$), summing the two doubles the real part and cancels the imaginary part, and we'll be left with a cosine wave at the appropriate phase:

$$z_m[n] + z_{N-m}[n] = A \cdot N \cdot \cos\left(2\pi \cdot \frac{m}{N} \cdot n + \phi\right)$$

This is exactly what we started with ($x[n]$), except there's an additional scaling of N in front. So, for at least this simple case of a single sinusoid, we can recover the input signal by generating the corresponding sinusoids with the parameters encoded by the magnitude and phase in the DFT, and dividing out the length of the signal N.

As we'll see in the next section, this strategy works in general: we don't need to assume that $x[n]$ is a single sinusoid!

7.2   THE INVERSE DFT

Generalizing the strategy used in the previous section's example, we get the following definition for an **inverse discrete Fourier transform** (IDFT).

Definition 7.1 (The Inverse DFT). Let $x[n]$ be an arbitrary signal of N samples, and let $X[m]$ be its DFT. The **inverse DFT** (IDFT) is defined as

$$x[n] = \frac{1}{N} \sum_{m=0}^{N-1} X[m] \cdot e^{+2\pi \cdot \mathrm{j} \cdot \frac{m}{N} \cdot n} \tag{7.1}$$

Intuitively, this says that the nth sample of the signal $x[n]$ can be recovered by averaging the nth samples of all DFT sinusoids.

Before proving the correctness of this definition, we should highlight the three key ways that it differs from the forward DFT defined by equation (5.12):

1. There is a global scaling of $1/N$;

2. The sign of the complex exponent is flipped: positive for inverse transform, negative for the forward transform;

3. The summation ranges over m (frequencies), rather than n (samples). Note that the *number* of frequencies (and samples) is still N, so the summation still ranges from $m = 0$ to $m = N - 1$.

The proof of DFT invertibility makes no assumptions about $x[n]$, apart from its duration (N).

Proof. Plugging in the definition of the DFT $X[m]$ (but using sample index n' to avoid confusion with n), we get the following:

$$\frac{1}{N} \sum_{m=0}^{N-1} X[m] \cdot e^{+2\pi \cdot \mathrm{j} \cdot \frac{m}{N} \cdot n} = \frac{1}{N} \sum_{m=0}^{N-1} \left(\sum_{n'=0}^{N-1} x[n'] \cdot e^{-2\pi \cdot \mathrm{j} \cdot \frac{m}{N} \cdot n'} \right) \cdot e^{+2\pi \cdot \mathrm{j} \cdot \frac{m}{N} \cdot n} \quad \text{DFT definition}$$

$$= \frac{1}{N} \sum_{n'=0}^{N-1} x[n'] \cdot \sum_{m=0}^{N-1} e^{-2\pi \cdot \mathrm{j} \cdot \frac{m}{N} \cdot n'} \cdot e^{+2\pi \cdot \mathrm{j} \cdot \frac{m}{N} \cdot n} \quad \text{Rearrange sum}$$

$$= \frac{1}{N} \sum_{n'=0}^{N-1} x[n'] \cdot \sum_{m=0}^{N-1} e^{-2\pi \cdot \mathrm{j} \cdot \frac{m}{N} \cdot n' + 2\pi \cdot \mathrm{j} \cdot \frac{m}{N} \cdot n} \quad e^a \cdot e^b = e^{a+b}$$

$$= \frac{1}{N} \sum_{n'=0}^{N-1} x[n'] \cdot \sum_{m=0}^{N-1} e^{2\pi \cdot \mathrm{j} \cdot \frac{n-n'}{N} \cdot m}$$

Now, there are two cases to consider. If $n' = n$, then the inner summation simplifies:

$$\sum_{m=0}^{N-1} e^{2\pi \cdot \mathrm{j} \cdot \frac{n-n'}{N} \cdot m} = \sum_{m=0}^{N-1} e^{2\pi \cdot \mathrm{j} \cdot \frac{0}{N} \cdot m} = \sum_{m=0}^{N-1} 1 = N \qquad \text{if } n' = n.$$

If $n' \neq n$, then the inner summation cancels totals to zero. This is because n' and n are both integers, and we can use the result of Section 5.7, except now with $n - n'$ taking the place of the frequency index, and m taking the place of the sample position:

$$\sum_{m=0}^{N-1} e^{2\pi \cdot \mathrm{j} \cdot \frac{n-n'}{N} \cdot m} = 0 \qquad \text{if } n - n' \neq 0.$$

The entire summation, therefore, has $N - 1$ terms contributing 0 and one term contributing $x[n] \cdot N$. Combining these cases, we can finish the derivation above:

$$\frac{1}{N} \sum_{n'=0}^{N-1} x[n'] \cdot \sum_{m=0}^{N-1} e^{2\pi \cdot \mathrm{j} \cdot \frac{n-n'}{N} \cdot m} = \frac{1}{N} \cdot x[n] \cdot N = x[n].$$

This is exactly what we needed to show: the nth sample is recovered exactly. □

7.2.1 The IDFT in practice

Like the forward DFT, the inverse DFT (IDFT) is implemented by most signal processing packages.

In Python, we have two ways to invert a DFT, depending on whether we have the full spectrum or only the real part:

```
# Full spectrum, all N analysis frequencies
X = np.fft.fft(x)

# Full inverse, should produce x_inv == x
x_inv = np.fft.ifft(X)

# Real-part only, 1 + N//2 analysis frequencies
Xr = np.fft.rfft(x)

# Real-part inverse, again produces x_inv == x
x_inv = np.fft.irfft(Xr)
```

7.2.2 Discussion

Nowhere in the proof of the inverse DFT did we assume anything about the signal contents $x[n]$: it works for **any signal** x. The entire derivation relies on the definition of the forward transform coefficients $X[m]$, and a couple of observations about summing complex sinusoids. So what does this actually tell us about the DFT?

The inverse DFT gives us an alternative representation of signals: every signal $x[n]$ can be uniquely represented as a combination of sinusoids:

- The summation in the inverse DFT $\sum_{m=0}^{N-1}$ represents the "combination";

- The coefficient $X[m]$ encodes the amplitude and phase of the mth sinusoid;

- The complex exponential $e^{2\pi \cdot \mathrm{j} \cdot m \cdot n / N}$ represents the mth sinusoid itself.

Up until this point, we've occasionally had to assume that such a representation exists. **But now we've proven that it exists**.

Aside from analysis and theoretical properties, the inverse DFT gives us tools to *modify* signals. Rather than operating on individual samples, we can alter the DFT coefficients to produce desired effects, and then take the inverse DFT to recover the time-domain signal. We'll have more to say about the frequency domain view of filtering in later chapters, but in the next section, we'll see how to use this insight for synthesizing signals directly.

7.3 SYNTHESIS

In the first section of this chapter, we saw how a signal $x[n]$ representing a sinusoid at an analysis frequency can be reconstructed from its DFT coefficients $X[m]$. In the second section, we saw (abstractly) that this is true of **any** signal $x[n]$, not just sinusoids. In this section, we'll unpack this a bit and look more carefully at some examples to get a better intuition for how this works.

The key idea is that the IDFT can be thought of as an example of **additive synthesis**: the signal $x[n]$ is reconstructed as a combination of pure tones.

7.3.1 IDFT as synthesis

Equation (7.1) tells us how to construct the nth sample of a signal x as an average of the following terms:

$$x[n] = \frac{1}{N} \sum_{m=0}^{N-1} X[m] \cdot e^{2\pi \cdot \mathrm{j} \cdot \frac{m}{N} \cdot n}$$

Remember that each DFT coefficient $X[m]$ is a complex number with an amplitude (we'll call it A_m) and a phase (ϕ_m):

$$X[m] = A_m \cdot e^{\mathrm{j} \cdot \phi_m}$$

This means that each term in IDFT summation can be equivalently expressed as

$$A_m \cdot e^{\mathrm{j} \cdot \phi_m} \cdot e^{2\pi \cdot \mathrm{j} \cdot \frac{m}{N} \cdot n} = A_m \cdot e^{\mathrm{j} \cdot \left(2\pi \cdot \frac{m}{N} \cdot n + \phi_m\right)}.$$

Equivalently, we can use Euler's formula to express this as waves instead of an exponential:

$$A_m \cdot \cos\left(2\pi \cdot \frac{m}{N} \cdot n + \phi_m\right) + \mathrm{j} \cdot A_m \cdot \sin\left(2\pi \cdot \frac{m}{N} \cdot n + \phi_m\right).$$

This says that each term in the summation has a real part (the cosine term) and an imaginary part (the sine term), but they both share the same amplitude A_m, frequency m/N, and phase ϕ_m.

As we saw in our *warmup example* earlier in the chapter, if the DFT has conjugate symmetry (so that $X[m] = \overline{X[N-m]}$), then the imaginary term corresponding to $X[m]$ will cancel with the imaginary term corresponding to $X[N-m]$ when we sum over all frequencies. This results in an entirely real-valued summation, where each term is produced by a sinusoid in standard form:

$$x[n] = \frac{1}{N} \sum_{m=0}^{N-1} A_m \cdot \cos\left(2\pi \cdot \frac{m}{N} \cdot n + \phi_m\right). \tag{7.2}$$

Note. If you've been wondering why we defined the standard form of sinusoids (1.3) by using cosine instead of sine, (7.2) is why. We could have equivalently written this equation using sine, but there would be extra phase offsets of $\pi/2$ to keep track of.

Note that this *only* works if the spectrum is conjugate symmetric. If we don't have conjugate symmetry, then the reconstructed signal may have complex values.

7.3.2 Example: partial reconstruction

The inverse DFT equation tells us how to reconstruct a signal from its spectrum, but we can also gain an intuition for how this works by looking at what happens when we stop the summation early, including only some $N' < N$ frequencies.

To visualize this process, we'll revisit the *trumpet example* from earlier. To keep the visualization simple, we'll take only a small fragment of the recording for analysis.

```
# We'll use soundfile to load the signal
import soundfile as sf

# And IPython to play it back in the browser
from IPython.display import Audio, display

# A single note (D#5) played on a trumpet
# https://freesound.org/s/48224/
# License: CC BY-NC 3.0
x, fs = sf.read('48224__slothrop__trumpetf3.wav')

# Slice out 512 samples (approximately 11ms) of audio
x = x[int(0.3 * fs):int(0.3*fs)+512]
```

Plotting the time-domain signal, along with its magnitude spectrum results in Fig. 7.2.

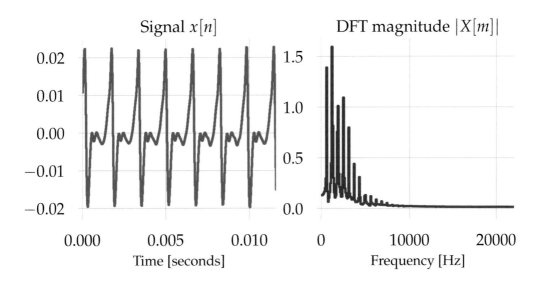

Figure 7.2 *Left*: Approximately 11ms (512/44100) excerpt of a trumpet playing D#. *Right*: The DFT magnitude spectrum of the signal (positive frequencies only).

We can see that the signal contains a repeating pattern, but also that it does not closely resemble a sinusoid. The DFT spectrum (Fig. 7.2, right) contains many peaks, indicating that the signal does indeed contain a rich combination of sinusoids. Fig. 7.3 below demonstrates how the signal would be approximated if only part of the DFT spectrum is used in the reconstruction.

A careful look at Fig. 7.3 reveals that the reconstruction converges fairly quickly to a good approximation of the original signal $x[n]$. While the later terms in the summation (corresponding to frequencies above 5000 Hz) contribute only small amounts to the sum, they are critical to capturing the beginning and end of the signal. This is because the signal we've excerpted is discontinuous at the boundaries, and as we saw in the discussion of *spectral leakage*, discontinuities require high-frequency content to accurately represent.

Finally, it is worth remembering that the visualization above does not illustrate the *phase* ϕ_m of the sinusoids being used in the reconstruction. This is primarily to keep the visualization simple: the phase content is still important to accurate reconstruction of the signal!

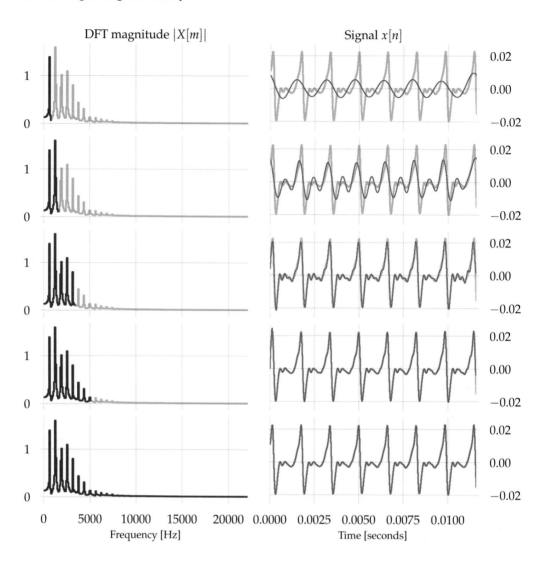

Figure 7.3 As more DFT coefficients $X[m]$ are included in the reconstruction (*left*), the reconstruction (*right*) better approximates the original input signal $x[n]$. Note here that only the positive frequencies ($m = 0, 1, \ldots, N/2$) are visualized here, and that each partial reconstruction up to frequency index m contains both m and $N-m$.

7.4 EXERCISES

Exercise 7.1. Synthesize a sinusoid $x[n] = A \cdot \cos(2\pi \cdot f \cdot n/N + \phi)$ by directly constructing the frequency domain representation $X[m]$ and applying the inverse DFT. Use the following starter code:

```
# sampling rate
fs = 8000
```

(continues on next page)

(continued from previous page)

```
# duration / number of frequencies
N = 8000

# Wave parameters
A = 5
f = 250   # Hz
phi = np.pi / 4

X = np.zeros(N, dtype=np.complex)

# INSERT YOUR CODE HERE TO BUILD X

# Invert to recover the signal
x = np.fft.ifft(X)
```

Hint. Check your synthesis against a signal generated purely in the time-domain. Your IDFT should be numerically identical to the time-domain signal, including amplitude and phase.

Make sure to construct your DFT so that the output is real-valued.

Exercise 7.2. A *low-pass filter* allows frequencies below some cut-off f_T to propagate through, and blocks all frequencies above the cut-off. Try implementing this on a signal of your choice by setting any DFT coefficients $X[m]$ to 0 if the corresponding frequency is above 500 Hz. Apply the inverse DFT and listen to the results. How does it sound? Do you notice any artifacts?

Hint. It is best to try this on a few different sounds, including natural and synthetic signals. Hand claps might be particularly interesting.

Exercise 7.3. Using a signal x of your choice (ideally a single pitched note or isolated sound), compute its DFT:

```
X = np.fft.fft(x)
```

and the following reconstructions:

- The inverse DFT: x_inv = np.fft.ifft(X).

- The inverse DFT with the original magnitudes and all phases set to 0: x_nophase = np.fft.ifft(np.abs(X)).

- The inverse DFT with the original phases and all magnitudes set to 1: x_nomag = np.fft.ifft(np.exp(1j * np.angle(X))).

Then, answer the following questions:

1. How do these three signals compare to each other?

2. How do they compare to the original signal?

3. What does this tell you about the relative importance of magnitude and phase?

4. How might your answers change with a different kind of test signal?

## CHAPTER 8

Fast Fourier transform

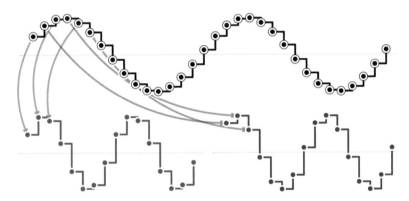

In this brief chapter, we'll see one of the most common methods for accelerating the computation of the discrete Fourier transform (DFT). The term "Fast Fourier Transform" (FFT) is often applied colloquially to mean any algorithm which computes the DFT more efficiently than the naive algorithm we described in *Chapter 5*. Because FFTs are so common in practice, completely supplanting naive DFT implementations, many people use the terms "FFT" and "DFT" interchangeably, but they are not the same thing!

Tip. An FFT is a fast algorithm for computing the DFT.

It is not a separate transform!

More specifically, in this chapter, we will see the so-called "radix-2 Cooley-Tukey" algorithm.

Before that, we'll first need to clearly state what we mean by *efficiency*, and analyze the naive DFT method to get a sense of what the baseline is.

8.1 TIME-ANALYSIS OF THE DFT

As we've seen in previous chapters, the *naive method* for computing the DFT uses two nested loops:

```
1   def dft(x):
2       '''Compute the Discrete Fourier Transform of an input signal x␣
        ↪of N samples.'''
3
4       N = len(x)
5       X = np.zeros(N, dtype=np.complex)
6
7       # Compute each X[m]
8       for m in range(N):
9
10          # Compute similarity between x and the m'th basis
11          for n in range(N):
12              X[m] = X[m] + x[n] * np.exp(-2j * np.pi * m * n / N)
13
14      return X
```

The outer loop (starting at line 8) iterates over the N different analysis frequencies: $m = 0, 1, 2, \ldots, N - 1$.

Within each of these loop is another loop (starting at line 11) that iterates over the *samples* to compute similarity between the input signal x and the reference complex sinusoid $e^{-2\pi \cdot j \cdot m \cdot n / N}$. Each step of this inner loop requires a constant amount of work doing basic arithmetic: multiplying, dividing, adding, and storing numbers. The exact amount of work here isn't too important: what matters is that it's the same amount for every choice of m and n.

Since there are N steps in the outer loop, and N steps inside each of those, line 12 is executed N^2 times in total. This means that as the length of the input grows (i.e., more samples are added), the amount of work grows *quadratically*, as shown below, in Figure 8.1.

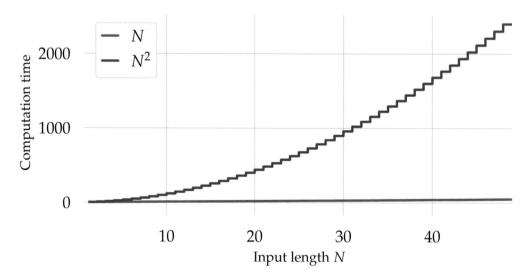

Figure 8.1 A comparison of linear (N) and quadratic (N^2) growth.

This quadratic growth in computation time establishes a baseline for DFT computation: the naive algorithm takes time proportional to N^2. Fast Fourier Transform methods aim to reduce this to scale more favorably as the input length increases.

Coming from the other direction, any algorithm for computing the DFT must take *at least* N steps, because it must see all N samples at least once, e.g., to compute $X[0] = \sum_n x[n]$. This gives us a lower bound on the computation time, and the goal will be to get as close to this lower bound as possible.

8.2 RADIX-2 COOLEY-TUKEY

As mentioned in the introduction to this chapter, there are many algorithms which are collectively referred to as "Fast Fourier Transforms". In this section, we'll see one of the earliest methods, (re-)discovered in 1965 by Cooley and Tukey [CT65], which can accelerate DFT calculations when N is an integral power of 2: $N = 2^K$.

> The Cooley-Tukey method for DFT calculation was known to Gauss all the way back in the early 19th century.
> What a clever chap.

Modern FFT implementations use many tricks to speed up calculation, and generalize to arbitrary values for N. That said, the core idea of the "radix-2 Cooley-Tukey" method has plenty of interest to offer, and sheds some light on how more advanced techniques can be developed.

8.2.1 Divide and Conquer

We'll start again with the DFT equation:

$$X[m] = \sum_{n=0}^{N-1} x[n] \cdot e^{-2\pi \cdot \mathrm{j} \cdot m \cdot n / N}.$$

The key observation of Cooley and Tukey is that this summation can be broken apart in interesting ways. Specifically, we can separate the summation into *even indices* $n = 0, 2, 4, \ldots$ and odd indices $n = 1, 3, 5, \ldots$ as shown in Figure 8.2.

Writing $n = 2k$ for even and $n = 2k + 1$ for odd indices, we get:

$$X[m] = \underbrace{\sum_{k=0}^{N/2-1} x[2k] \cdot e^{-2\pi \cdot \mathrm{j} \cdot m \cdot 2k / N}}_{\text{Even part}} + \underbrace{\sum_{k=0}^{N/2-1} x[2k+1] \cdot e^{-2\pi \cdot \mathrm{j} \cdot m \cdot (2k+1) / N}}_{\text{Odd part}}$$

All we've done so far is change variables: the summations are identical to the original definition. However, if we notice that $2k/N = k/(N/2)$, we can interpret the first summation as the DFT $X_E[m]$ where $x_E[n] = x[2n]$ is the signal comprised of the even-numbered samples:

$$\sum_{k=0}^{N/2-1} x[2k] \cdot e^{-2\pi \cdot \mathrm{j} \cdot m \cdot 2k / N} = \sum_{k=0}^{N/2-1} x[2k] \cdot e^{-2\pi \cdot \mathrm{j} \cdot m \cdot k / (N/2)}$$

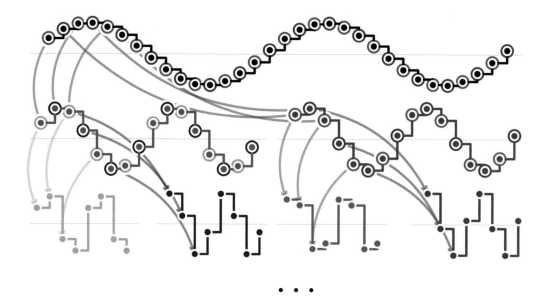

Figure 8.2 A discrete signal $x[n]$ of $N = 32$ samples (top row) is divided into its *even* and *odd* samples (middle-left and middle-right). The even and odd signals themselves can be similarly divided (bottom row). This process can repeat up to $\log_2 N$ times.

This trick doesn't exactly work for the odd-part summation, but we can make it work with a little arithmetic. Observe that by the rule of exponents $e^{a+b} = e^a \cdot e^b$,

$$e^{-2\pi \cdot \mathrm{j} \cdot m \cdot (2k+1)/N} = \left(e^{-2\pi \cdot \mathrm{j} \cdot m \cdot 2k/N} \right) \cdot \left(e^{-2\pi \cdot \mathrm{j} \cdot m/N} \right).$$

The first factor is exactly the same as in the even case, and gives us the DFT basis. The second factor, $e^{-2\pi \cdot \mathrm{j} \cdot m/N}$ (commonly called a *"twiddle factor"*) does not depend on the variable of summation (k), and can therefore be factored out of the sum, giving us:

$$\sum_{k=0}^{N/2-1} x[2k+1] \cdot e^{-2\pi \cdot \mathrm{j} \cdot m \cdot (2k+1)/N} = \left(e^{-2\pi \cdot \mathrm{j} \cdot m/N} \right) \cdot \sum_{k=0}^{N/2-1} x[2k+1] \cdot e^{-2\pi \cdot \mathrm{j} \cdot m \cdot k/(N/2)}$$

This says that the second summation is the DFT $X_O[m]$ of the signal comprised of odd-indexed samples $x_O[n] = x[2n + 1]$, with an additional rotation of $-2\pi \cdot m/N$.

Combining these two observations, we can express the original DFT of N samples in terms of two DFTs of $N/2$ samples:

$$X[m] = X_E[m] + \left(e^{-2\pi \cdot \mathrm{j} \cdot m/N} \right) \cdot X_O[m] \qquad (8.1)$$

> Note that the multiplicative factor $e^{-2\pi \cdot \mathrm{j} \cdot m/N}$ is exactly what we would get from the *DFT shifting theorem* when x_O is delayed by one sample.

This expression is an example of the **divide-and-conquer** approach to problem solving. When given a large problem to solve, it sometimes helps to break it into smaller problems whose solutions can be combined to solve the original problem.

Why is this helpful?

So far, all we've done is express a single component of the DFT $X[m]$ *recursively* in terms of two smaller DFT's, $X_E[m]$ and $X_O[m]$. If we only cared about computing a single frequency index m, this recursive formulation wouldn't help us. Remember, we can compute a single DFT coefficient $X[m]$ in N steps (the inner loop of the naive algorithm in the previous section), and we know that it can't be done in fewer than N steps because each sample $x[n]$ must be looked at.

Of course, we're usually interested in computing the full DFT (all N frequencies). The recursive decomposition of $X[m]$ is helpful here because it applies simultaneously to *all frequencies* $m = 0, 1, 2, \ldots, N - 1$. That is, if someone gave us the entire DFT calculations for X_E and X_O, we could compute the entire DFT X in N additional steps by applying (8.1) (once for each $m = 0, 1, 2, \ldots, N - 1$). In principle, we should be able to do something like the following:

> **Warning.** The pseudo-code here does not quite work for the reasons given below. But it should give you a sense of how smaller DFTs can be combined to produce the full DFT.

```python
def pseudo_fft(x, X_E, X_O):
    '''Compute the DFT of x given the even and odd DFTs X_E, X_O'''

    N = len(x)

    # Combine the given even and odd spectra
    X = np.zeros(N, dtype=np.complex)

    for m in range(N):
        X[m] = X_E[m] + np.exp(-2j * np.pi * m / N) * X_O[m]

    return X
```

In reality, we won't get X_E and X_O for free. However, if the total work taken to get X_E and X_O is less than the naive algorithm's N^2, this still amounts to an improvement.

8.2.2 How many frequencies?

A careful look at (8.1) reveals something a little strange: the DFT $X[m]$ involves N frequencies ($m = 0, 1, 2, \ldots, N - 1$), while the smaller DFTs $X_E[m]$ and $X_O[m]$ should have only $N/2$ frequencies ($m = 0, 1, 2, \ldots, N/2 - 1$) since they are derived from signals of length $N/2$. This introduces a problem for lines 9–10 in the code block above: how do we recover all N frequencies when only $N/2$ are present in X_E and X_O?

Once again, **aliasing** can come to the rescue.

Recall that each $X[m]$ is the result of a comparison to a wave that completes m

cycles in N samples, or in units of Hz:

$$f_m = \frac{m}{N} \cdot f_s.$$

For (8.1) to make sense, $X_E[m]$ should involve a comparison to a wave that complete m cycles in $N' = N/2$ samples. Note, however, that the sampling rate for the even signal x_E is $f'_s = f_s/2$ because the period between samples is double that of our original signal x. If $m \geq N/2$, the aliasing equation (2.3) tells us that this frequency has an aliasing frequency f_a given by:

$$
\begin{aligned}
f_a &= \frac{m}{N'} \cdot f'_s - f'_s \\
&= \frac{m}{N/2} \cdot \frac{f_s}{2} - \frac{f_s}{2} \\
&= \frac{m}{N} \cdot f_s - \frac{f_s}{2} \\
&= \frac{m - N/2}{N} \cdot f_s.
\end{aligned}
$$

In plain words, this says that to find $X_E[m]$ for $m \geq N/2$, we should look at the DFT coefficient $X_E[m - N/2]$. This is relationship is depicted visually in Fig. 8.3.

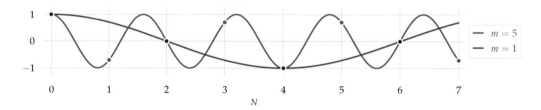

Figure 8.3 A wave at $m = N/2 + 1$ (with $N = 8$, $m = 5$) is decimated by taking only the even-index samples $n = 0, 2, 4, \ldots$.. Due to aliasing, the resulting sequence is equivalent to a wave at $m - N/2 = 1$.

Frequency indices $m < N/2$ do not need special aliasing treatment, as they will occur exactly where we expect them to in the DFT spectrum: $X_E[m]$.

We can notationally combine these two cases ($< N/2$ and $\geq N/2$) as

$$m \rightarrow \left(m \mod \frac{N}{2} \right),$$

or, in Python code:

```
m % (N //2)
```

This allows us to correctly state the expression of $X[m]$ in terms of X_E and X_O for all m:

$$X[m] = X_E \left[m \mod \frac{N}{2} \right] + \left(e^{-2\pi \cdot j \cdot m/N} \right) \cdot X_O \left[m \mod \frac{N}{2} \right].$$

The benefit of this form is that we only rely on $N/2$ coefficients in both X_E and X_O, but we are still able to recover all N coefficients $X[m]$. This, ultimately, is the source of the efficiency gains in FFT algorithms: the solutions to smaller sub-problems can be used multiple times to produce different parts of the output. $X[m]$ and $X[m+N/2]$ both depend on $X_E[m]$ and $X_O[m]$. We compute X_E and X_O once, but we get to use them twice.

8.2.3 A radix-2 algorithm

Finally, combining the observations above, we can give a full definition of the radix-2 FFT algorithm.

The last ingredient of this algorithm, compared to the naive DFT implementation, is *recursion*. To compute X_E and X_O, the algorithm will **call itself** on the appropriately sub-sampled signal. Once it has X_E and X_O, it then combines the results and returns the output spectrum.

> **Recursion is weird.**
>
> If you've never seen recursion before, don't panic.
>
> Yes, it's weird, but it's not magic.
>
> If it helps, think of a simpler recursive algorithm for adding up a list of numbers. You can think of any such sum as "the first number, plus whatever the sum of the rest is". That definition also applies to the "sum of the rest" part, and every subsequent sum until there is only a single number left.

If the input has only one sample ($N = 2^0 = 1$), then there is only one analysis frequency ($m = 0$), and $X[0] = x[0]$ is just the sample value itself. This represents the **base case**, the input which terminates the recursion.

```python
def fft2(x):
    '''Compute the DFT of an input x of N = 2**k samples'''

    N = len(x)  # Get the length of the input

    # The DFT of a single sample is just the sample value itself
    if N == 1:
        return x

    else:
        # Recursively compute the even and odd DFTs
        X_even = fft2(x[0::2])  # Start at 0 with steps of 2
        X_odd = fft2(x[1::2])   # Start at 1 with steps of 2

        X = np.zeros(N, dtype=np.complex)  # Allocate the output
        ↪array

```

(continues on next page)

(continued from previous page)

```
17        # Combine the even and odd parts
18        for m in range(N):
19            # Find the alias of frequency m in the smaller DFTs
20            m_alias = m % (N//2)
21            X[m] = X_even[m_alias] + np.exp(-2j * np.pi * m / N) * X_
↪odd[m_alias]
22
23        return X
```

Time analysis

Analyzing the running time of the `fft2` algorithm above requires a slightly more sophisticated approach than the naive DFT algorithm, due to the use of recursion.

Let $T(N)$ represent the amount of time taken by the algorithm on an input of length N. For now, $T(N)$ is an unknown quantity, but this notation will help us solve for it. We're not going to aim for a precise count of the operations performed by the algorithm, but merely a coarse upper bound on how the computation scales with the length of the input (N).

The base case, $T(1)$, takes some constant amount of time, which we can call c. This accounts for all computation up to line 10 where it returns from the $N = 1$ case.

If $N > 1$, then we see two recursive calls to inputs of length $N/2$. This is followed by a loop of N steps (line 21), of which each step takes some constant amount of time that we'll call d, for a total of $N \cdot d$. (We'll absorb the amount of time to allocate the output array at line 18 into this quantity.) Using this observation, we can express the total time for $T(N)$ recursively:

$$T(N) = 2 \cdot T\left(\frac{N}{2}\right) + N \cdot d$$

Each of these recursive calls, in turn, has two more recursive calls plus $d \cdot N/2$ work to combine the results:

$$T(N) = 2 \cdot \left(2 \cdot T\left(\frac{N}{4}\right) + \frac{N}{2} \cdot d\right) + N \cdot d$$

and this process can repeat until the recursion stops at $N = 1$.

If we total up the non-recursive computation done at each *level* of the recursion, we can get a handle on the total amount of computation. This is illustrated in Fig. 8.4. To simplify notation, we'll assume that $c \leq d$, meaning that the arithmetic operations at lines 23–24 are more expensive than the base-case operations in lines 1–10.

What Fig. 8.4 shows us is that each level ℓ of the recursion has 2^ℓ calls to `fft2`, each of size $N/2^\ell$. Each of these does some linear amount of computation: $d \cdot N/2^\ell$, and if we sum these together, the result for level ℓ is $d \cdot N$. The total number of levels is $1 + \log_2 N$, since this counts how many times N can be divided by 2 before hitting the base case: $N = 1$ has one level, $N = 2$ should have two levels, $N = 4$ has 3 levels,

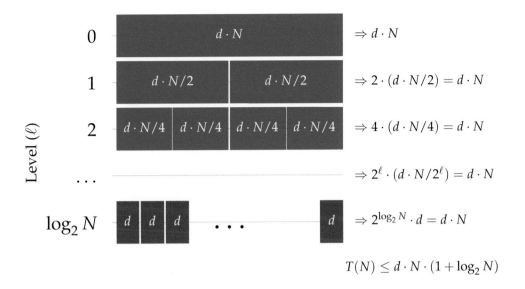

Figure 8.4 The total amount of computation performed by the radix-2 FFT algorithm (`fft2`) can be computed by looking at the non-recursive computation done at each level, and then adding up the levels. Each level totals at most $d \cdot N$ computation, and there are $1 + \log_2 N$ levels.

and so forth. Multiplying the per-level work by the number of levels gives us a total upper bound of

$$T(N) \leq d \cdot N \cdot (1 + \log_2 N).$$

Since $\log_2 N < N$, this comes out to be substantially more efficient than the $d \cdot N^2$ work performed by the naive approach.

> In asymptotic algorithm analysis, we typically ignore constant factors and focus on the largest term that scales with the input length. Since $1 + \log_2 N < 2 \cdot \log_2 N$, this means that we would ignore any linear dependencies on N.
>
> The common notation for this is $T(N) \in \mathcal{O}(N \cdot \log N)$, which we read as "$T(N)$ is big-Oh (order of) $N \cdot \log N$".

8.2.4 Summary

In this chapter, we've seen that the radix-2 Cooley-Tukey algorithm is substantially faster than the naive DFT implementation. This is especially true when N is large, as is the common case in audio signal processing.

The code given above for `fft2` is a rather simplistic implementation given to illustrate the basic concepts. It can be made much more efficient in several ways, including: pre-computing and caching the "twiddle" factors, re-using a single output buffer rather than re-allocating arrays for each partial output, and so on.

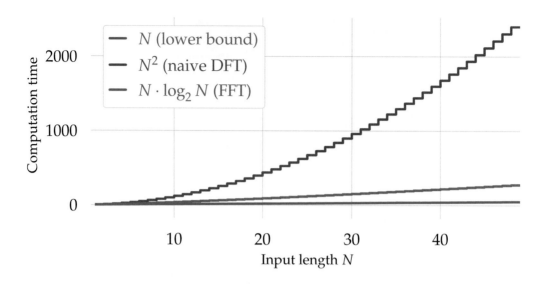

Figure 8.5 A comparison of linear (N), quadratic (N^2), and *linearithmic* $(N \cdot \log_2 N)$ growth. The linearithmic curve is much gentler than the quadratic curve, meaning that the amount of work increases slowly with the size of the input.

The main limitation of the radix-2 method is that it only works if N is an integral power of 2: $N = 1, 2, 4, 8, 16, \ldots$. If $N = 37$ (for example), this method cannot be used.

The radix-2 method is just one special case of the general method of Cooley and Tukey. In the radix-2 case, we divide an input of length N into 2 inputs of length $N/2$. More generally, if N is divisible by some integer p, we can divide into p inputs of length N/p. The basic principle behind this more general "mixed-radix" approach is the same: the DFTs of the smaller cases are combined to form the larger case by applying the appropriate delay ("twiddle factor") to each one. This more general approach retains the $N \log N$ computational complexity for broader classes of input length (not just powers of 2).

Where the Cooley-Tukey method fails is when the input length N is a prime number (eg, 37, or 257) and cannot be divided evenly into pieces. In these cases, alternate methods have been developed which still achieve running time that scales like $N \log N$.

In practice, modern FFT implementations – such as the Fastest Fourier Transform in the West (FFTW) – use many combinations of strategies to optimize the computation time for a given input length.

The take-home message is that computing DFTs can be done in much less than quadratic time:

Tip. The DFT for a signal of N samples can be computed (by FFT methods) in time $\approx N \cdot \log_2 N$.

8.3 EXERCISES

Exercise 8.1. Imagine that you want to optimize the radix-2 method to only compute the positive frequencies (i.e., like `np.fft.rfft`). How would you change the method?

Exercise 8.2. If you have an input $N = 3^k$ (for some integer k), how would you modify the radix-2 algorithm to get a radix-3 algorithm?

1. How many recursive calls will the FFT function need to make?

2. What should the sub-problems represent, if not "even"- and "odd"-indexed signals?

3. How would you define the "twiddle" factors for combining the results of recursive calls?

## CHAPTER 9

The short-time Fourier transform

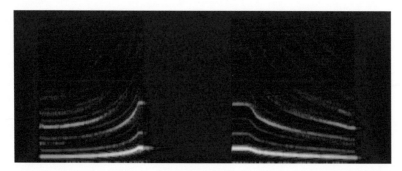

The Fourier transform, as presented in earlier chapters, provides a view of signals in terms of different frequencies. The comparison between the input signal and the Fourier basis involves a sample-wise product and sum over *all samples* to produce a single DFT component $X[m]$. Implicitly, this is assuming that frequency content is somehow "stationary" or unchanging over the duration of the signal. While this may be a reasonable assumption for short signals that loop indefinitely, it does not apply to reasonably long signals, whose frequency content may be expected to change over time. (For instance, speech or music!)

In this chapter, we will see how to relax the assumption of stationarity, providing a **time-frequency** representation of signals that analyzes frequency content only over short intervals. This so-called *Short-Time Fourier transform* (STFT) is a fundamental tool in analyzing dynamic signals, and has a long history of use in speech processing [AR77, FG66].

9.1   FRAMING

The key idea behind the Short-Time Fourier transform is to divide a long signal up into short pieces, and analyze each piece separately. While the frequency content of a signal may change over long periods of time, real signals tend to be approximately stationary if we look only at short fragments.

DOI: 10.1201/9781003264859-9

> **Frames**
>
>   "Frame" is one of those unfortunate terms in digital signal processing that has multiple meanings, depending on context. This can be confusing for newcomers!
>
>   The definition we'll use is that a "frame" corresponds to a relatively small number of audio samples, e.g., an array of 50 samples:
>
> $$[x[100], x[101], x[102], \ldots, x[149]].$$
>
> In other sources, you may also encounter the terms **buffer** or **window** for this concept. We use **frame** here to distinguish from the more general concept of a *buffer* as temporary storage, and the concept of a *window* defined in *earlier chapters*.
>
>   The alternate definition of **frame** that you may see elsewhere defines a "frame" for a multi-channel signal (e.g., a stereo mix with left- and right-channels) as the array of sample values for each channel at a specific sample index, e.g., the length-2 array $[x_L[1000], x_R[1000]]$.

These fragments are known as **frames**, and typically span a few thousand samples at a time. The number of samples in a frame is conveniently known as the **frame length**, which we'll denote by N_F. The frame length determines the duration of the fragment being analyzed ($t = N_F/f_s$).

In addition to the frame length, we also must decide how many frames to take. This is controlled by the **hop length**: the number of samples between frames, which we will denote by N_H.

The framing process is illustrated in Fig. 9.1.

Combining these two quantities, the nth sample of the kth frame is given by

$$y[k, n] = x[k \cdot N_H + n] \qquad \text{For } n = 0, 1, \ldots, N_F - 1$$

Here, we are using a two-dimensional array y to represent the *framed signal*: the first index k selects which frame, and the second index n selects which sample within the frame.

The kth frame, therefore, is the slice of the signal from sample indices $k \cdot N_H$ to $k \cdot N_H + N_F - 1$.

Equivalently, in Python, the kth frame would by computed by the array slice:

```
x[k * NH:k * NH + NF]
```

Using some dimensional analysis, we can convert frame indices to time as well:

$$(k \, [\text{frames}]) \cdot \left(N_H \frac{[\text{samples}]}{[\text{frame}]} \right) \cdot \left(t_s \frac{[\text{seconds}]}{[\text{sample}]} \right) = k \cdot N_H \cdot t_s \, [\text{seconds}]$$

$$= k \cdot \frac{N_H}{f_s} \, [\text{seconds}].$$

From this, we can observe that N_H/f_s can be interpreted as the (time) period between frames. Equivalently, its reciprocal f_s/N_H gives us the number of frames per second, a quantity known as the **frame rate**.

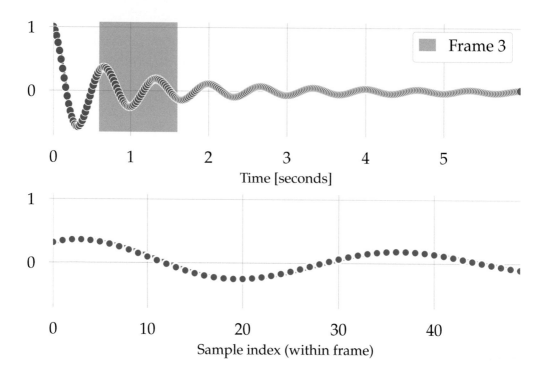

Figure 9.1 *Top*: a signal $x[n]$ of $N = 300$ samples taken at $f_s = 100$. *Bottom*: the frames constructed from the signal with frame length $N_F = 50$ and hop length $N_H = 10$.

9.1.1 How many frames are there?

Calculating the exact number of frames in a signal from the framing parameters can be a little subtle, but it's not too hard if we're careful.

It can be helpful to think of a few of extreme cases:

1. What if $N_F = N$, so that the frame length is exactly the same as the signal length? In this case, there should be only 1 frame. The hop length does not matter here because any index offset other than 0 would push the frame off the end of the input array, and we would not have a full frame.

2. What if $N_F = N_H = 1$? In this case, each sample is a frame by itself, so we should have N frames (one per sample).

3. What if $N_F = 1$ and $N_H = 2$? In this case, we're effectively decimating the signal by a factor of 2 (taking every other sample), so we should have $N/2$ frames (if N is even) or $(N-1)/2$ (if N is odd).

More generally, how many steps k of size N_H can we take before $k \cdot N_H + N_F \geq N$? Rearranging this inequality gives us the following formula.

Note (Counting frames). A signal of N samples, with frame length $N_F \leq N$ and hop length N_H will produce K frames, where

$$K = 1 + \left\lfloor \frac{N - N_F}{N_H} \right\rfloor. \tag{9.1}$$

The "extra" $1+$ comes from the fact that frame $k = 0$ does not invoke a step by the hop length; if we did not have this extra 1, the result would not agree with the "extreme" cases identified above.

Note that (9.1) divides the signal length (minus one frame) by the hop length. However, the hop length generally may not evenly divide $N - N_F$, which is why we round this ratio down so that only **full frames** are counted.

The Python equivalent of (9.1) is:

```
# Integer division // rounds down
N_frames = 1 + (len(x) - N_F) // N_H
```

Now that we have a handle on how to apply framing to a signal, we are ready to define the Short-Time Fourier transform.

9.2 DEFINING THE STFT

The Short-Time Fourier Transform (STFT) does exactly what it says: it applies the Fourier transform to short fragments of time, that is, frames taken from a longer signal. At a conceptual level, there is not too much going on here: we just extract frames from the signal, and apply the DFT to each frame. However, there is much to discuss in the details.

9.2.1 A basic STFT algorithm

A basic STFT algorithm requires three things:

- the input signal x,

- the frame length N_F, and

- the hop length N_H.

Typical STFT implementations assume a real-valued input signal, and keep only the non-negative frequencies by using `rfft` instead of `fft`. The result is a two-dimensional array, where one dimension indexes the frames, and the other indexes frequencies. Note that the frame length dictates the number of samples going into the DFT, so the number of analysis frequencies will also be N_F.

Warning. This is probably not the STFT code you want to use in practice. This implementation is intentionally simplified to illustrate the basic idea.

```
def basic_stft(x, n_frame, n_hop):
    '''Compute a basic Short-Time Fourier transform
    of a real-valued input signal.'''

    # Compute the number of frames
    frame_count = 1 + (len(x) - n_frame) // n_hop

    # Initialize the output array
    # We have frame_count frames
    #     and (1 + n_frame//2) frequencies for each frame
    stft = np.zeros((frame_count, 1 + n_frame // 2), dtype=np.
↪complex)

    # Populate each frame's DFT results
    for k in range(frame_count):
        # Slice the k'th frame
        x_frame = x[k * n_hop:k * n_hop + n_frame]

        # Take the DFT (non-negative frequencies only)
        stft[k, :] = np.fft.rfft(x_frame)

    return stft
```

Fig. 9.2 demonstrates the operation of this `basic_stft` method on a real audio recording.

The type of visualization used in Fig. 9.2 may look familiar to you, as it can be found on all kinds of commercially available devices (stereos, music software, etc.). Now you know how it works.

9.2.2 Spectrograms

Another way of representing the output of a Short-Time Fourier transform is by using **spectrograms**. Spectrograms are essentially an image representation of the STFT, constructed by stacking the frames horizontally, so that time can be read left-to-right, and frequency can be read bottom-to-top. Typically, when we refer to *spectrograms*, what we actually mean are *magnitude spectrograms*, where the phase component has been discarded and only the DFT magnitudes are retained. In Python code, we would say:

```
# Compute the STFT with frame length = 1024, hop length = 512
stft = basic_stft(x, 1024, 512)

# Take the absolute value to discard phase information
S = np.abs(stft)
```

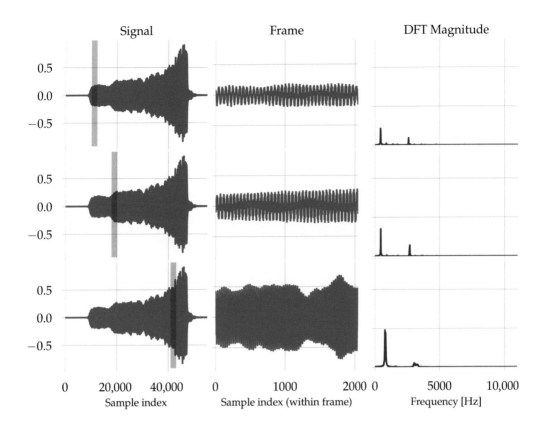

Figure 9.2 A signal x is sampled at $f_s = 22{,}050$ and frames are taken with $N_F = 2048$ and $N_H = 512$. Each frame of $x[n]$ (*left*, shaded region) is plotted (*middle*) along with its DFT magnitudes $|X[m]|$ as produced by the STFT (*right*).

This allows us to interpret energy $(S = |X|)$ visually as brightness under a suitable color mapping.

Fig. 9.3 (top) illustrates an example of a spectrogram display. Each column (vertical slice) of the image corresponds to one frame of Fig. 9.2 (right).

While some spectral content is visually perceptible in Fig. 9.3 (top), most of the image is dark, and it's generally difficult to read. This goes back our earlier discussion of *decibels*: human perception of amplitude is logarithmic, not linear, so we should account for this when visualizing spectral content.

The bottom plot of Fig. 9.3 shows the same data, but using a decibel scaling for amplitudes:

$$S_{\mathrm{dB}} = 20 \cdot \log_{10} S$$

The result of this mapping exposes far more structure in the input signal. The (framewise) fundamental frequency of the signal is visually salient as the bright contour at the bottom of the image, but harmonics are also visible, as is background noise.

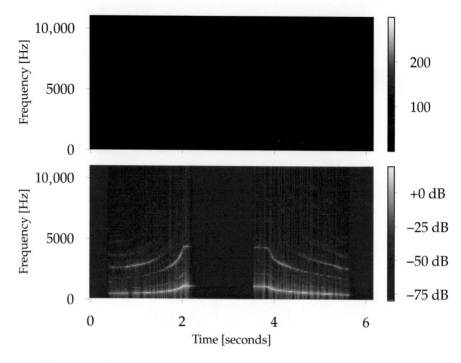

Figure 9.3 The *magnitude spectrogram* representation of the slide whistle example of Fig. 9.2, using the same parameters $N_F = 1024$, $N_H = 512$. **Top**: visualization using linear magnitude scaling. **Bottom**: visualization using decibel scaling.

Calculating dB

The decibel calculation in Fig. 9.3 is actually a bit more detailed.

First, the spectrogram magnitudes are normalized by $S \to S/\sqrt{N_F}$. This ensures that spectrograms of different frame lengths will have comparable magnitude scales. This normalization does not change the visual characteristics of the plot, only the interpretation of the colors.

Second, to avoid numerical underflow when taking the log of small numbers (close to 0), a bias offset of 10^{-4} is added. This ensures that the dB measurement is always at least

$$20 \cdot \log_{10} 10^{-4} = -80,$$

which prevents the visual scale from being skewed by inaudible artifacts.

The total dB calculation is more accurately represented as

$$S_{\mathrm{dB}} = 20 \cdot \log_{10}\left(S/\sqrt{N_F} + 10^{-4}\right).$$

9.2.3 Choosing parameters

The `basic_stft` algorithm above has two parameters that we are free to set however we see fit. There is no single "right" setting for these STFT parameters, but there are settings that will be better or worse for certain applications.

Frame length N_F

Unlike the standard DFT, where the number of analysis frequencies is dictated by the number of samples, the STFT allows us to control this parameter directly. This introduces a **time-frequency** trade-off.

Large values of N_F will provide a high-frequency resolution, dividing the frequency range $[0, f_s/2]$ into smaller pieces as N_F increases. This comes at the cost of reduced time resolution: large values of N_F integrate over longer windows of time, so any changes in frequency content that are shorter than the frame length could be obscured. Intuitively, when the frame length is large (and the hop length is fixed), any given sample $x[n]$ will be covered by more frames, and therefore contribute to more columns in the spectrogram, resulting in a blurring over time.

Conversely, small values of N_F provide good time localization–since each frame only sees a small amount of information–but poor frequency resolution, since the range $[0, f_s/2]$ is divided into relatively few pieces.

Fig. 9.4 visualizes this trade-off for a fixed hop length N_H and varying frame length N_F.

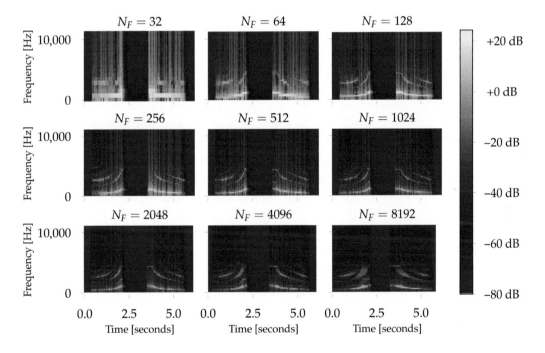

Figure 9.4 Changing the frame length $N_F = 32, 64, 128, \ldots, 8192$ illustrates the time-frequency trade-off. At low frame lengths, energy spreads out vertically along the frequency axis; at large frame lengths, it spreads horizontally along the time axis. For intermediate values of N_F, this spread is relatively balanced, providing a clear representation of the time-varying frequency content of the signal. All analyses here use $N_H = 512$ and $f_s = 22{,}050$.

As a general rule, it is common to choose N_F to be an integral power of 2 (e.g., 1024, 2048, etc.). This is almost entirely done for efficiency purposes, as it allows for the use of the *radix-2 FFT* algorithm. However, this is not a *requirement* of the STFT: any frame length $N_F > 0$ will work.

Tip. In typical audio applications, absent any other constraints, one heuristic for choosing the frame length is to ensure that it is long enough to observe at least a few cycles at the low end of the audible frequency range (which have the longest periods: $t_0 = 1/f_0$).

If we assume that humans can perceive frequency down to 20 Hz, this frequency has period $1/20 = 0.05$ [seconds], so two cycles would take 0.1 seconds. If the sampling rate is $f_s = 44,100$ (so that the Nyquist frequency is above the upper limit of audible frequencies of 20,000 [Hz]), then a frame length $N_F = 4096$ would have duration $N_F/f_s \approx 0.093$ [seconds], which is pretty close to 0.1.

Note that this is just a heuristic, and not a rule: other choices for N_F may be more appropriate for different situations, but this line of reasoning can lead you to a reasonable starting place.

Hop length N_H

The hop length parameter behaves a bit differently from the frame length parameter, as it has no bearing on the frequency resolution of the STFT. All it controls is how often we compute the DFT, so an STFT with a large hop length can be understood as a decimation (in time) of an STFT with a small hop length.

Small values of the hop length produce highly redundant outputs, meaning that the (magnitude) spectrogram at frame k is similar to that of $k+1$ if N_H is small. This makes sense if we're assuming that frequency content is locally consistent and not changing too rapidly. Redundancy isn't necessarily a bad thing, but it does require more memory and more processing: remember, the number of frames in an STFT scales like N/N_H (ignoring constants and boundary effects).

Larger values of hop length provide coarser time resolution. Note, however, that if $N_H > N_F$ (that is, we hop by more than a frame length each time), then **some samples will not be covered by any frame**. This would potentially lose information, especially due to transients that fall between frames. It is commonly recommended, therefore, to set N_H as a constant fraction of N_F, typically either 1/8, 1/4, or 1/2.

Frame overlap

Many STFT implementations (and reference texts) do not expose the hop length parameter directly, but require it to be set as a fraction of the frame length. For example, you may see STFT implementations like `scipy.signal.stft`, that take an `overlap` parameter instead of a hop length.

We have decided here to describe the STFT in terms of the hop length parameter for two reasons. First, it avoids any ambiguity due to rounding when multiplying the overlap ratio by the frame length. Second, it more naturally connects to how a dynamic STFT would be implemented when an audio stream is captured in real time.

This ensures that no samples are lost in the STFT, and due to overlapping frames, the resulting STFT should vary smoothly.

Fig. 9.5 illustrates the effects of varying hop length for a fixed frame length.

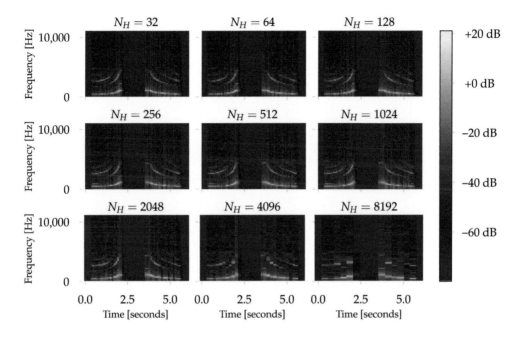

Figure 9.5 Changing the hop length $N_H = 32, 64, 128, \ldots, 8192$. At small hop lengths, we have a dense sampling in time of the frequency content, and a high-resolution image. At large hop lengths, the time resolution is decreased because frames are taken less frequently. However, the frequency resolution remains the same. All analyses here use $N_F = 2048$ and $f_s = 22{,}050$.

9.2.4 Windowing

An astute observer might notice in Figs. 9.4 and 9.5 that there are vertical banding artifacts in the spectrograms, which suggest broad-band bursts of energy across the entire frequency spectrum which are typically associated with transient events, e.g., percussion or other abrupt discontinuities in the signal. However, the input signal consists of a continuously varying pitch played on a slide whistle with no discernible transients. What's going on?

These artifacts are due to *spectral leakage*: for a fixed frame length N_F, it's unlikely that a period signal will line up exactly with the frame boundaries. As we saw perviously, the DFT will use sinusoids of all available analysis frequencies to explain the abrupt discontinuity that would occur if the frame was looped indefinitely, and this results in a distorted view of the signal through the STFT.

We can resolve this by using *windowing*: sample-wise multiplying the framed signal by a curve that tapers to 0 at the beginning and end so that the looped frame appears continuous.

> The **window** goes in the **frame**. Get it?

It is common for STFT implementations to provide this feature, as illustrated below by `wstft`:

```python
def wstft(x, n_frame, n_hop, window):
    '''Compute a windowed Short-Time Fourier transform

    Parameters
    ----------
    x : the input signal (real-valued)
    n_frame : the frame length (in samples)
    n_hop : the hop length (in samples)
    window : a window specification
        e.g., "hann" or ("kaiser", 4.0)
        See scipy.signal.get_window for details

    Returns
    -------
    stft : the Short-Time fourier transform of x (complex-valued)
        Shape: (frame_count, 1 + n_hop // 2)
    '''

    # Compute the number of frames
    frame_count = 1 + (len(x) - n_frame) // n_hop

    # Initialize the output array
    # We have frame_count frames and (1 + n_frame//2) frequencies␣
↪for each frame
    X = np.zeros((frame_count, 1 + n_frame // 2), dtype=np.complex)

    # We'll use scipy's window constructor here
    window = scipy.signal.get_window(window, n_frame)

    # Populate each frame's DFT results
    for k in range(frame_count):
        # Slice the k'th frame, apply a window, and take its DFT
        X[k, :] = np.fft.rfft(x[k * n_hop:k * n_hop + n_frame] *␣
↪window)

    return X
```

To use the windowed STFT, one must provide the choice of window (typically by name), for example:

```
# Use a Hann window
stft = wstft(x, 2048, 512, 'hann')

# A 'rectangular' window is equivalent to no window at all
stft_nowin = wstft(x, 2048, 512, 'rectangular')
```

The implementation above uses scipy.signal.get_window to construct the specified window, and many choices are available.

Fig. 9.6 illustrates the difference between a Hann-windowed and non-windowed STFT.

Tip. Always use windowing in your STFT, unless you have a strong reason not to.

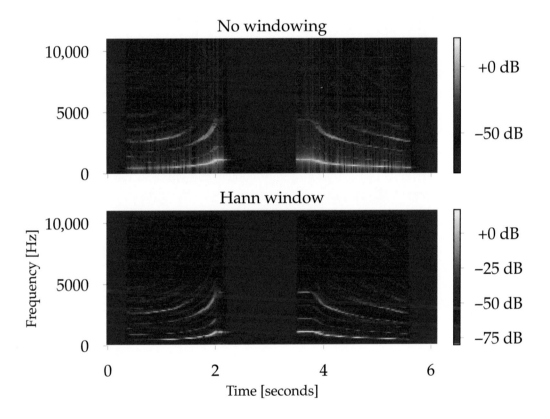

Figure 9.6 **Top**: An STFT computed using basic_stft without any windowing. **Bottom**: A windowed STFT computed by wstft with window='hann'. The windowed STFT has significantly less spectral leakage, which reduces the vertical banding artifacts, and makes harmonic structure more visible. Both analyses use $N_F = 2048$, $N_H = 512$, and $f_s = 22{,}050$.

9.2.5 Summary

The STFT provides a way to apply Fourier analysis locally to small fragments of a long signal, thereby relaxing the assumption that frequency content is stationary over the duration of the signal. It therefore has become a standard starting point for all kinds of signal processing tasks, especially those involving real-time signal acquisition. While it may look a bit complicated at first, with experience, people can learn to "read" spectrogram visualizations and identify properties and contents of an audio signal by visual inspection alone.

9.3   EXERCISES

Exercise 9.1. Imagine you have a signal with length $N = 44,100$ sampled at $f_s = 44,100$. How many frames would you get if you take an STFT with frame length $N_F = 4096$ and hop length $N_H = 512$?

Exercise 9.2. Sometimes, one does not want to discard any samples when performing an STFT. This can be done by **padding** the signal with trailing zeros. In the configuration from question 1, what is the smallest number of samples that you would need to add to capture the entire signal?

Can you give a more general form for calculating the required padding, in terms of (unknown) parameters N, N_F, N_H?

Exercise 9.3. The SciPy package provides an STFT implementation `scipy.signal.stft` which uses a slightly different parametrization that the one presented in this chapter.

Using a (non-trivial) test signal x of your choice, can you find parameter settings of `scipy.signal.stft` that produce identical output to the `wstft` function for $N_F = 2048$ and $N_H = 512$?

Hint. It's easiest to check the shape of the outputs first:

```
import numpy as np
import scipy

# [COPY in wstft definition from the text]

s1 = wstft(x, n_frame, n_hop, 'hann')
s2 = scipy.signal.stft(...)

# Check shapes
assert s1.shape == s2.shape
```

Note: you may need to transpose `s2` by saying `s2 = s2.T` so that the time- and frequency dimensions are in the same order as ours.

After you get the shapes to line up, test for numerical equivalence by using `np.allclose`

```
assert np.allclose(s1, s2)
```

## CHAPTER 10

Frequency domain convolution

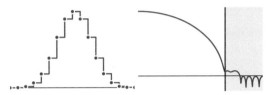

The third part of this book focuses primarily on **filtering**: how do we manipulate signals, and how do we reason about these manipulations?

In this chapter, we'll see how to use the discrete Fourier transform to understand *convolution*.

10.1   THE CONVOLUTION THEOREM

Recall from (3.1) the definition of **convolution** between a signal x (of length N) and impulse response h (of length K):

$$y[n] = (h * x)[n] = \sum_{k=0}^{K-1} h[k] \cdot x[n-k].$$

As we saw previously, this can be interpreted as constructing the output y by mixing together several delayed copies of x (index k corresponds to the delay in samples), each with a gain coefficient given by $h[k]$.

In this section, our goal is to understand the frequency domain representation $Y = \mathrm{DFT}(h * x)$ in terms of the DFTs of the inputs h and x, which will be expressed succinctly by the **convolution theorem**.

10.1.1   Circular convolution

Before we can properly apply the DFT to convolution, we'll first need to deal with the differing assumptions of how negative sample indices are handled. When we first saw convolution, it was defined using the assumption that $x[n-k] = 0$ if $n < k$, i.e., that the signal x is silent before the 0th sample. Different ways of interpreting this assumption gave rise to the different *convolution modes* (full, valid, and same).

The DFT, on the other hand, assumes that signals *repeat* indefinitely, so that $x[n-k] = x[n-k+N]$.

DOI: 10.1201/9781003264859-10

If we define convolution using the repetition assumption, we get what is known as **circular convolution**. The equation is exactly the same as (3.1); all that has changed is the interpretation of negative sample indices, which now wrap around to the end of the signal. This assumption can be encoded by using modular arithmetic to compute the delayed sample index: $n - k \mod N$.

Definition 10.1 (Circular convolution). For a signal x of length N and impulse response h of length K, the **circular convolution** between x and h is defined as:

$$y[n] = \sum_{k=0}^{K-1} h[k] \cdot x[n - k \mod N].$$

10.1.2   The convolution theorem

Now that we've defined circular convolution, we can formally state the **convolution theorem**, which is one of the most important theorems in signal processing.

Theorem 10.1 (The Convolution Theorem). :label: conv-thm Let h and x be sequences of length N, and let $y = h * x$ denote the circular convolution between them.

The DFT of the convolution is the product of the DFTs:

$$y = h * x \qquad \Leftrightarrow \qquad Y[m] = H[m] \cdot X[m]. \tag{10.1}$$

Proof. By definition, the output signal y is a sum of delayed copies of the input $x[n - k]$, each scaled by the corresponding coefficient $h[k]$.

By *DFT linearity*, we can think of the DFT $Y[m]$ as a weighted combination of DFTs:

$$Y = \text{DFT}(y) = \text{DFT}\left(\sum_{k=0}^{N-1} h[k] \cdot x[n - k] \right) \qquad \text{By definition of } y$$

$$= \sum_{k=0}^{N-1} h[k] \cdot \text{DFT}(x[n - k]) \qquad \text{By DFT linearity.}$$

To reduce clutter, let's define $X_k = \text{DFT}(x[n - k])$ to be the DFT of the k-step delayed copy of x.

Now, we can use the *DFT Shifting Theorem* to express the DFT of the delayed signal X_k in terms of the original signal $X = \text{DFT}(x)$ for each frequency index m:

$$X_k[m] = X[m] \cdot e^{-2\pi \cdot j \cdot m \cdot k / N}.$$

Substituting this into our derivation for Y, we can continue:

$$Y[m] = \sum_{k=0}^{N-1} h[k] \cdot X_k[m]$$

$$= \sum_{k=0}^{N-1} h[k] \cdot X[m] \cdot e^{-2\pi \cdot j \cdot m \cdot k / N} \qquad \text{DFT Shifting theorem}$$

$$= X[m] \cdot \sum_{k=0}^{N-1} h[k] \cdot e^{-2\pi \cdot j \cdot m \cdot k / N} \qquad X[m] \text{ does not depend on } k$$

$$= X[m] \cdot H[m] \qquad \text{Definition of DFT of } h$$

$$= H[m] \cdot X[m] \qquad \text{By commutativity: } w \cdot z = z \cdot w.$$

which is exactly what we set out to prove. □

10.1.3 Consequences of the convolution theorem

Now that we have the convolution theorem, let's take some time to explore what it gives us.

Fast convolution

From the convolution theorem, we get $Y[m] = H[m] \cdot X[m]$. Applying the inverse DFT, we can recover the time-domain output signal:

$$y[n] = \text{IDFT}(H \cdot X)$$

where the product $H \cdot X$ is the element-wise product of H and X.

This gives us a recipe for computing the convolution $h * x$ using the DFT:

1. Pad x and h to the same length (if necessary)

2. Compute DFTs X and H,

3. Multiply $Y[m] = H[m] \cdot X[m]$,

4. Take the inverse DFT of $y = \text{IDFT}(Y)$.

If h and x are of comparable length (say, both N samples), then this can be more efficient than the *direct convolution algorithm*. The direct convolution has an outer loop of N steps (one for each output sample) and an inner loop of N steps (one for each delay), for a total of $\sim N^2$ computation.

However, the recipe above takes:

1. $\leq N$ steps for padding

2. Two DFT calculations

3. N steps to multiply the spectra

4. One inverse DFT calculation

Steps 2 and 4 can be performed with $\sim N + N \cdot \log N$ steps. The total computation time will therefore scale like $N \cdot \log N \ll N^2$.

However, if h is much shorter than N, this may not be worth it, as the direct method would take $N \cdot K$ steps. Standard convolution implementations like `scipy.signal.convolve` typically compare the lengths of the signal to determine the most efficient means of computing the result.

Algebraic properties

You may recall from the earlier section on *properties of convolution* that we asserted (without proof) that convolution is commutative:

$$h * x = x * h,$$

and associative:

$$h * (g * x) = (h * g) * x,$$

but we did not prove those properties. (We could have, but it would have been pretty tedious.)

The convolution theorem provides a more direct and intuitive way to see these properties:

- Commutativity must exist because complex multiplication is commutative: $X[m] \cdot H[m] = H[m] \cdot X[m]$.

- Likewise for associativity: $H[m] \cdot (G[m] \cdot X[m]) = (H[m] \cdot G[m]) \cdot X[m]$.

Inheriting these properties from multiplication is much easier than deriving them from scratch.

Filter analysis

Finally, the convolution theorem provides a way to understand the effect a particular impulse response h might have on a signal. Initially, we thought of each $h[k]$ as the gain applied to the delayed signal $x[n - k]$.

In the frequency domain, each $H[m]$ is a complex number, which we can denote as

$$H[m] = A \cdot e^{j \cdot \phi},$$

where $A \geq 0$ is the magnitude and ϕ is the phase.

Using the inverse DFT, we can interpret $X[m] = B \cdot e^{j \cdot \theta}$ as encoding the magnitude B and phase θ of a particular sinusoid present in the signal x.

Multiplying these two together, we see that $Y[m]$ has magnitude $A \cdot B$ and phase $\phi + \theta$. That is, $H[m]$ encodes how much each sinusoidal component of x is amplified (or attenuated) (by A) and delayed (by ϕ) when convolved with h.

In the following sections, we'll see how to put this idea into practice for filtering signals.

Dual convolution theorem

One interesting corollary of the convolution theorem is the following:

Corollary 10.1 (Convolution in frequency). If w and x are sequences of length N, then element-wise multiplication in the time domain is equivalent to circular convolution *in the frequency domain.*

$$\text{DFT}(w \cdot x) = \frac{1}{N} \cdot \text{DFT}(w) * \text{DFT}(x),$$

where frequency-domain convolution is defined as follows with $W = \text{DFT}(w)$ and $X = \text{DFT}(x)$:

$$(X * W)[m] = \sum_{k=0}^{N-1} W[m] \cdot X[m - k \quad \text{mod } N]$$

The proof of *Corollary 10.1* is nearly identical to that of the convolution theorem, except that it uses a variation of the shifting theorem for the *inverse* DFT.

The dual convolution theorem is mainly useful as a theoretical device, as it can help us to understand the effects of element-wise multiplication. This occurs when windowing a signal, and if we apply the inverse DFT to each side of *Corollary 10.1*, we obtain

$$w \cdot x = \frac{1}{N} \cdot \text{IDFT}(W * X).$$

We won't go into detail here, but the table below summarizes the relationships between DFT, IDFT, circular convolution, and element-wise multiplication.

Time domain		Frequency domain
Convolution $h * x$	$\Leftrightarrow$	Multiplication $H \cdot X$
Multiplication $w \cdot x$	$\Leftrightarrow$	Convolution $W * X$

10.2 CONVOLUTIONAL FILTERING

In the previous section, we saw that the convolution theorem lets us reason about the effects of an impulse response H in terms of each sinusoidal component. Specifically, the convolution theorem says that if $y = h * x$, then

$$Y[m] = H[m] \cdot X[m],$$

and we can interpret $H[m] = A \cdot e^{j \cdot \phi}$ as applying a gain A and delay ϕ to the sinusoid of frequency $\frac{m}{N} \cdot f_s$.

We'll investigate delay and phase in the next section, and focus here on the gain. Intuitively, if the magnitude A is small,

$$A = |H[m]| \approx 0,$$

then $|Y[m]|$ must be small as well, regardless of $X[m]$. This gives us a way to attenuate or eliminate frequencies in X when constructing Y!

Alternately, if A is large, then the corresponding frequency will be amplified. This section will focus primarily on the attenuation case, which often has more practical relevance.

10.2.1   Low-pass filters

To begin, we'll create a **low-pass filter**, meaning that the filter allows low frequencies to pass through to the output, and stops high frequencies.

Why low-pass filters?

We focus on low-pass filters here because A) they're useful, and B) they can be used to construct other filters.

If we apply a low-pass filter to x resulting in y, then $x - y$ should contain whatever is left over: this gives us a **high-pass** filter.

If we then apply another low-pass filter (with a higher cut-off frequency than before), we'd get a **band-pass** filter.

If we subtract the result of the band-pass filter from the original signal, we'd get a **notch** filter.

Moving average filters

One way to create a low-pass filter (though not a good one) is to average sample values together over a short window of time:

$$y[n] = \sum_{k=0}^{K-1} \frac{1}{K} \cdot x[n-k],$$

which is known as a **moving average filter**. We can define the impulse response of this filter as

$$h[n] = \begin{cases} \frac{1}{K} & n < K \\ 0 & \text{otherwise,} \end{cases}$$

and the length K is known as the *order* of the filter.

As illustrated below, high frequencies will oscillate within the time window, and average to zero (or close to it), while low frequencies will not.

The moving average can be understood as an average of delay filters:

$$h = \frac{1}{K} \cdot ([1, 0, 0 \ldots] + [0, 1, 0, 0, \ldots] + \ldots),$$

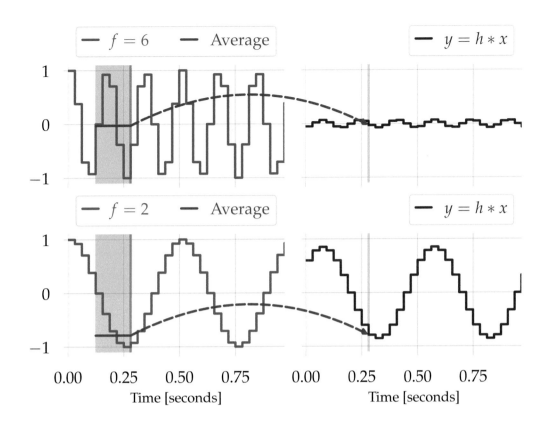

Figure 10.1 **Left**: two waves at frequencies $f = 6$ (top) and $f = 2$ (bottom) are sampled at $f_s = 32$ for one second. A moving average filter of order $K = 5$ (shaded region) is applied to produce the output signals $y = h * x$ via circular convolution (**right** plots). The solid horizontal lines (left) correspond to the average value over the window (marked points in the right plots). The high frequency signal (top) is significantly attenuated, while the lower frequency signal is not.

so the DFT H can be interpreted as the average of the DFT of delays. As we saw previously, the *DFT of a delay is a sinusoid*. By DFT linearity, the DFT of the average of delays is therefore the average of the DFT of delays. So if h is an order-K moving average filter, then its DFT will be

$$H[m] = \frac{1}{K} \cdot \sum_{k=0}^{K-1} e^{-2\pi \cdot \mathrm{j} \cdot m \cdot k / N},$$

as visualized below in Figure 10.2.

To understand the effect of this filter, we can focus on the spectral magnitudes $|H[m]|$, as these control how much each frequency will be attenuated.

As Fig. 10.3 shows, some frequencies are attenuated substantially, but the effect is not monotonic: the magnitude curve $|H|$ has *ripples*, and some higher frequencies are less attenuated than lower frequencies.

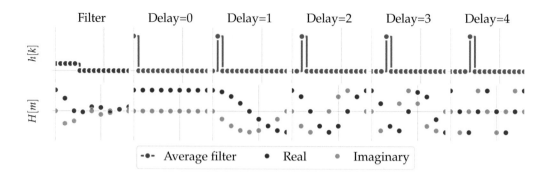

Figure 10.2 An order-5 moving average filter h (**top left**) can be constructed as the average of 5 delay filters (**top row**). Each delay's DFT is a sinusoid (**bottom row**), and their sample-wise average produces the DFT of the moving average (**bottom-left**).

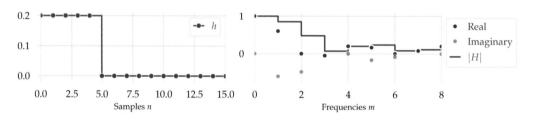

Figure 10.3 **Left**: an order 5 moving average filter h. **Right**: its DFT H (real and imaginary components), as well as its magnitude spectrum $|H|$.

This is why the moving average filter is not particularly good, and should not be used in practice! Note, however, that we learned this entirely by analyzing the filter itself through the lens of the convolution theorem.

10.2.2 The "ideal" low-pass filter

An alternative approach to building a filter would be to start in the frequency domain. If we want to get rid of all frequencies above some cut-off f_c, why not just set $H[m] = 0$ (if $f_m > f_c$) and $H[m] = 1$ otherwise? This way, low frequencies should be preserved exactly, and high frequencies will be completely eradicated, right?

This type of filter is known as an **ideal low-pass filter** (LPF) or a *brick-wall filter*. For example, if we want to apply an ideal low-pass filter $f \geq 5$ with $f_s = N = 32$ as in our running example, we could do the following in Python:

```
fs = 32
N = 32

# Initialize the filter as ones
H = np.ones(1 + N // 2, dtype=np.complex)
```

(continues on next page)

(continued from previous page)

```
# Get the analysis frequencies
freqs = np.fft.rfftfreq(N, d=1/fs)

# Set any frequencies above the cutoff to 0
cutoff = 5
H[freqs >= cutoff] = 0

# Invert to recover the time-domain filter
h = np.fft.irfft(H)
```

The result of this process is visualized below in Figure 10.4.

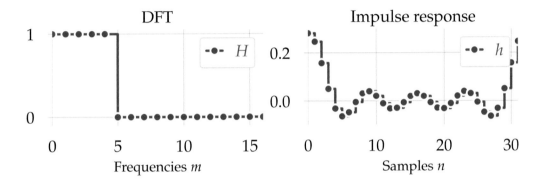

Figure 10.4 **Left**: an ideal low-pass filter H is constructed for $f_s = 32$, $N = 32$, and $f_c = 5$. **Right**: the time-domain filter is recovered by $h = \text{IDFT}(H)$.

Note that in Fig. 10.4, the impulse response h peaks at the beginning and end of the filter, and is relatively low in the middle. It can be easier to visualize if we circularly shift the filter so that it peaks in the center. This will change the delay characteristics of the filter, but not its frequency response.

This can be done manually by `np.fft.fftshift(h)`. Note that most FIR filter constructors, like those covered in the next section, will implement this automatically for you.

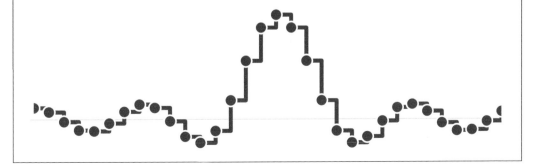

Fig. 10.4 illustrates the time-domain representation (impulse response) h of the "ideal" low-pass filter H constructed in the frequency domain. It may be surprising that h appears to be periodic, and oscillates around 0. However, this can be understood by reasoning about its inverse DFT as a sum of cosines (since the imaginary parts of H are all 0):

$$h[n] = \frac{1}{N} \cdot \sum_{m=-4}^{4} \cos\left(2\pi \cdot \frac{m}{N} \cdot n\right)$$

The range of the summation here is -4 to +4. Note that Fig. 10.4 shows only the non-negative frequencies. The terms $m = 0, \ldots, 4$ come from the non-negative frequencies, and terms $m = -4, \ldots, -1 \equiv N - 4, \cdots, N - 1$ come from DFT conjugate symmetry.

These oscillations produce audible artifacts, known as **ringing**, when h is convolved with a signal. The example below demonstrates this, using an audio clip of a slide whistle and an ideal low-pass filter at 500 Hz: ringing at the cut-off frequency is audible in the low-pass filtered output. For comparison purposes, a moving average filter is also included.

```python
import soundfile as sf
from IPython.display import Audio, display

# https://freesound.org/s/517633/
x, fs = sf.read('517633__samuelgremaud__slide-whistle-1-mono.wav')
N = len(x)

# Define the cutoff frequency
f_cutoff = 500

# For comparison, a moving average filter
# at fs [samples / second] and f_cutoff [cycles / second]
# we get an order K = fs / f_cutoff [samples / cycle]
K = fs // f_cutoff
h_ma = np.ones(K) / K

# Filter the signal: moving average
y_ma = scipy.signal.convolve(x, h_ma)

# Ideal LPF: set all DFT components above f_cutoff to 0
X = np.fft.rfft(x)
freqs = np.fft.rfftfreq(N, 1/fs)
X[freqs >= f_cutoff] = 0
y = np.fft.irfft(X)   # Invert the DFT to get the output
```

(continues on next page)

(continued from previous page)

```
display('Original signal')
display(Audio(data=x, rate=fs))

display(f'Moving average, cutoff = {f_cutoff} Hz')
display(Audio(data=y_ma, rate=fs))

display(f'Ideal LPF, cutoff = {f_cutoff} Hz')
display(Audio(data=y, rate=fs))
```

In principle, the ideal LPF ought to work, but its utility is limited in practice by the length of the filter, which determines the number of analysis frequencies. Just as sharp edges (e.g., impulses) in the time domain are difficult to represent by continuous sinusoids, the same is true for sharp edges in the *frequency domain*. The sharper the filter is in the frequency domain, the more frequencies (equivalently, time-domain samples) are necessary to represent it accurately. The ideal filter is infinitely sharp – it has a slope of $-\infty$ at the cutoff frequency – so it requires infinitely many samples to represent in the time domain. Any finite approximation we take, like the one above, will produce ringing artifacts.

Because of these ringing artifacts, ideal low-pass filters are almost never used in practice. However, they can be modified to produce better filters, as we will see in the next section.

10.3 FILTER DESIGN AND ANALYSIS

In the previous section, we saw first how the frequency domain view of convolutional filters lets us reason about their effects. We then saw that the ideal low-pass filter produced some less-than-ideal results when applied to real signals, notably ringing artifacts which arise from using finite-length approximations to the ideal filter.

In this section, we'll develop this idea further, and see how to construct better convolutional filters.

10.3.1 Terminology

Before we go further into filter design, it will help to establish some terminology.

For a low-pass filter with cutoff f_c, the **pass band** is the set of frequencies $0 \leq f \leq f_c$ that pass through the filter.

The set of frequencies which are blocked by the filter is referred to as the **stop band**.

In general, there is a region between the pass and stop bands where frequencies are attenuated, but still audible, which is known as the **transition band**.

These regions are illustrated below for an example filter in Fig. 10.5.

Within the pass-band, a filter may not have perfectly flat response. The amount of variation within the pass-band (max-min) is known as **pass-band ripple**. An

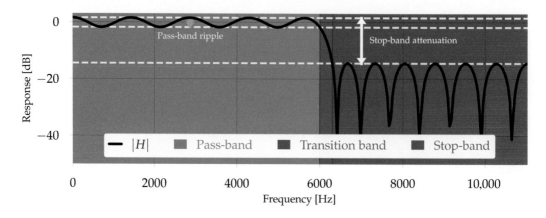

Figure 10.5 The *frequency response* plot of a filter h is the magnitude of its DFT H. This filter was designed to have a cutoff of $f_c = 6000$ Hz and a transition band of 300 Hz.

ideal filter would have 0 pass-band ripple, which amounts to no change in amplitude for passing frequencies.

Similarly, the difference from the peak response of the pass-band to the peak of the stop-band is the **stop-band attenuation**. This measures how audible the presumably stopped frequencies will be in the output. An ideal filter would have infinite attenuation, but the filter in Fig. 10.5 has only about 16dB, which is well within the range of human hearing. (This is probably not a good filter.)

What makes a good filter?

The diagram above gives us a way to compare different filters. In general, the desirable properties of a filter are:

1. High stop-band attenuation. *The filter stops frequencies when it should.*

2. Small pass-band ripple. *Passing frequencies aren't too distorted.*

3. Narrow transition band. *The filter efficiently moves from pass to stop behavior.*

These three properties often present a trade-off, and can depend on the order of the filter in complicated ways. Designing filters isn't easy, but this section will give you some pointers to standard methods.

10.3.2 The window method

One of the most common approaches to making low-pass filters is known as the *window method*, and it works by applying a window to the impulse response of the ideal low-pass filter. This is similar to the approach taken earlier to combat *spectral leakage*, though for a slightly different reason here. Rather than force the filter to be periodic, windowing here simply forces the filter to depend only on a finite number of samples, and reduces ringing artifacts.

The window method proceeds by first determining the order K of the filter, which is typically an integer p multiple of the cutoff frequency f_c (measured in samples):

$$K = \left\lfloor \frac{p \cdot f_s}{f_c} \right\rfloor \quad \text{[samples]}.$$

Next, an ideal LPF is constructed for length K and its impulse response h is computed by the inverse DFT. Finally, a window w (e.g., Hann) is constructed for length K and multiplied by h to produce the windowed filter:

$$h_w[k] = w[k] \cdot h[k].$$

This process is visually demonstrated by Fig. 10.6.

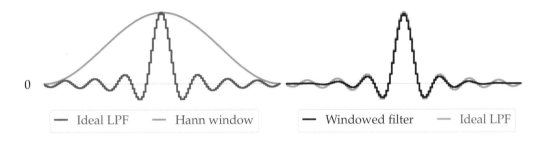

Figure 10.6 **Left**: the impulse response of an ideal low-pass filter is multiplied by a window function to produce the windowed low-pass filter (**right**).

The `scipy` package helpfully implements all of this (and more) for us in one place: `scipy.signal.firwin` (finite impulse response + window). For example, we could construct a Hann-windowed low-pass filter at $f_c = 500$ by the following code:

```
f_cutoff = 500

# Two cycles of the cutoff frequency (in samples)
order = 2 * fs // f_cutoff

# Build the filter
hw = scipy.signal.firwin(order, f_cutoff, window='hann', fs=fs)

# Apply the filter to a signal
y = scipy.signal.convolve(x, hw)
```

The examples below demonstrate this process for Hann-windowed filters with $p = 1, 2$, and 4 cycles of the cutoff frequency, and the filters are visualized in Fig. 10.7.

```
# Make the filter long enough to catch two cycles of our cutoff␣
 ↪frequency
order1 = 2 * fs // f_cutoff
hw1 = scipy.signal.firwin(order1, f_cutoff, window='hann', fs=fs)
y_hw1 = scipy.signal.convolve(x, hw1, mode='same')

# Or four cycles
order2 = 4 * fs // f_cutoff
hw2 = scipy.signal.firwin(order2, f_cutoff, window='hann', fs=fs)
y_hw2 = scipy.signal.convolve(x, hw2, mode='same')

# Or 8 cycles
order3 = 8 * fs // f_cutoff
hw3 = scipy.signal.firwin(order3, f_cutoff, window='hann', fs=fs)
y_hw3 = scipy.signal.convolve(x, hw3, mode='same')

# Or 16 cycles
order4 = 16 * fs // f_cutoff
hw4 = scipy.signal.firwin(order4, f_cutoff, window='hann', fs=fs)
y_hw4 = scipy.signal.convolve(x, hw4, mode='same')

display(f'Hann-windowed low-pass filter, order={order1}')
display(Audio(data=y_hw1, rate=fs))
display(f'Hann-windowed low-pass filter, order={order2}')
display(Audio(data=y_hw2, rate=fs))
display(f'Hann-windowed low-pass filter, order={order3}')
display(Audio(data=y_hw3, rate=fs))
display(f'Hann-windowed low-pass filter, order={order4}')
display(Audio(data=y_hw4, rate=fs))
```

From Fig. 10.7, we can see that taking longer filters (higher order) results in a better approximation to the ideal LPF, and the output of the filters in each case does not have the ringing artifacts of the ideal LPF.

scipy.signal.firwin can do much more than just low-pass filters. It can directly construct high-pass, band-pass, multi-band filters, and more.

When the cutoff frequency f_c is high, the window method can work well. However, when the cutoff frequency is low (like in this example) this method requires extremely long filters to achieve narrow transition bands.

10.3.3 The Parks-McClellan method

The window method is not the only way to create a filter. Probably the next most commonly used method is due to Parks and McClellan [PM72], and it attempts to minimize ripple in both the pass and stop bands. The algorithm is known alternately

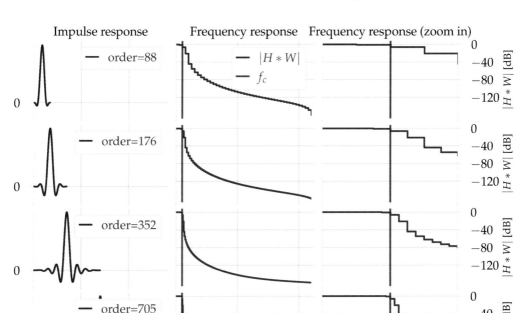

Figure 10.7 **Left**: Hann-windowed low-pass filters for $p = 2, 4, 8, 16$ cycles of the cutoff frequency $f_c = 500$ (with $f_s = 44{,}100$). **Middle**: the frequency domain representation of the windowed filters (decibel-scaled). **Right**: a zoomed-in view of the filter DFTs, covering the frequency range $[0, 1000]$ Hz. Note that as the filter length increases, so does the slope of the filter for frequencies above f_c, as well as the number of DFT coefficients.

as *Parks-McClellan* or *Remez*, the latter after Evgeny Remez, who developed the underlying method to solve the optimization problem several decades earlier [Rem34].

Unlike the window method, the Parks-McClellan method requires the user to explicitly describe the transition band, and the desired gain in each band (pass and stop). In Python, this can be done as follows:

```
f_cutoff = 500

# One cycle of the cutoff frequency
order = fs // f_cutoff

# Build the filter using the Remez method
# We'll use the following bands:
#    Pass band: [0, 0.75 * f_cutoff]  (begin to transition at 75%␣
↪of our cutoff)
```

(continues on next page)

(continued from previous page)

```
#             gain: 1
#     Stop band: [f_cutoff, fs/2]   (attenuate everything above the␣
↪cutoff)
#             gain: 0

hpm = scipy.signal.remez(order,
                        # Our frequency bands:
                        [0, 0.75 * f_cutoff, f_cutoff, fs/2],
                        # Our target gain for each band:
                        [1, 0],
                        # And sampling rate:
                        fs=fs)


# Apply the filter to a signal
y = scipy.signal.convolve(x, hpm)
```

This example starts the transition band before f_c, allowing a flat frequency response in the stop band as demonstrated below in Fig. 10.8. The following example outputs demonstrate this filter design applied to our example at multiple orders. Just as with the window method, higher orders are more precise but less computationally efficient.

```
# Make the filter long enough to catch two cycles of our cutoff␣
↪frequency
order1 = 2 * fs // f_cutoff
hpm1 = scipy.signal.remez(order1,
                        [0, 0.75 * f_cutoff, 1.0 * f_cutoff, fs/2],
                        [1., 0.0], fs=fs)
y_hpm1 = scipy.signal.convolve(x, hpm1, mode='same')

order2 = 4 * fs // f_cutoff   # Or four cycles
hpm2 = scipy.signal.remez(order2,
                        [0, 0.75 * f_cutoff, 1.0 * f_cutoff, fs/2],
                        [1., 0.0], fs=fs)
y_hpm2 = scipy.signal.convolve(x, hpm2, mode='same')

order3 = 8 * fs // f_cutoff   # Or 8 cycles
hpm3 = scipy.signal.remez(order3,
                        [0, 0.75 * f_cutoff, 1.0 * f_cutoff, fs/2],
                        [1., 0.0], fs=fs)
y_hpm3 = scipy.signal.convolve(x, hpm3, mode='same')

order4 = 16 * fs // f_cutoff   # Or 16 cycles
```

(continues on next page)

(continued from previous page)

```python
hpm4 = scipy.signal.remez(order4,
                          [0, 0.75 * f_cutoff, 1.0 * f_cutoff, fs/2],
                          [1., 0.0], fs=fs)
y_hpm4 = scipy.signal.convolve(x, hpm4, mode='same')

display(f'Parks-McClellan filter, order={order1}')
display(Audio(data=y_hpm1, rate=fs))
display(f'Parks-McClellan filter, order={order2}')
display(Audio(data=y_hpm2, rate=fs))
display(f'Parks-McClellan filter, order={order3}')
display(Audio(data=y_hpm3, rate=fs))
display(f'Parks-McClellan filter, order={order4}')
display(Audio(data=y_hpm4, rate=fs))
```

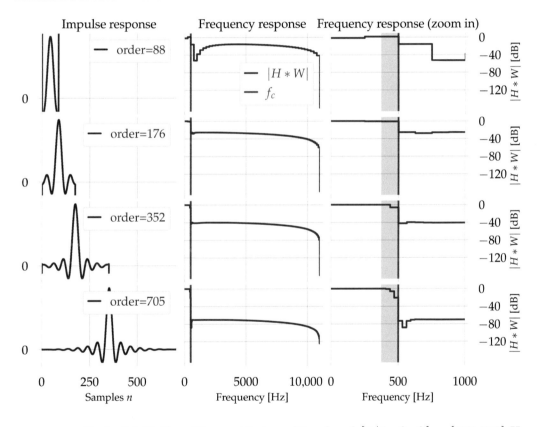

Figure 10.8 Parks-McClellan filters with transition band $[3/4 \cdot f_c, f_c] = [375, 500]$ Hz (shaded regions) at multiple orders ($p = 2, 4, 8, 16$ cycles of f_c). As in Fig. 10.7, increasing the order improves the attenuation of the filter.

Comparing the frequency responses of window-method filters Fig. 10.7 to those of the Parks-McClellan filters Fig. 10.8, we can see that the windowed filters have gentler slope, and the Parks-McClellan filters have a steeper transition with a flat response in the stop-band. Both are effective at suppressing high-frequency content, though neither method will completely remove energy at frequencies above f_c. That's okay though – usually all we care about when filtering is reducing the stop-band energy to the point where it is not perceptible.

10.4 PHASE AND GROUP DELAY

In the previous section, we saw how to investigate the properties of a convolutional filter h by examining its frequency response, that is, the magnitude of its DFT $|H|$. Examining the magnitude tells us an important part of the story, since it directly encapsulates the gain applied to each frequency when the filter is applied to a signal. However, it's not the entire story: **phase is also important**.

Recall that the convolution theorem converts time-domain convolution into frequency-domain multiplication:

$$y = h * x \qquad \Leftrightarrow \qquad Y[m] = H[m] \cdot X[m]$$

Since each $H[m]$ is a complex number, we can express it as $H[m] = A \cdot e^{j\phi}$ for magnitude A and phase ϕ. This tells us that the output DFT component $Y[m]$ will be derived from $X[m]$ by scaling (multiply by A) and rotation by angle ϕ:

$$Y[m] = H[m] \cdot X[m] = A \cdot e^{j\phi} \cdot X[m].$$

From the *DFT shifting theorem*, we know that applying delay to a signal results in a rotation of each DFT component $X[m]$, so we can interpret the rotation by ϕ as implementing a delay of the mth sinusoid. The shifting theorem states precisely how this works: a delay of k samples induces a rotation of $-2\pi \cdot k \cdot m/N$ for frequency index m with signal length N, so that each frequency is rotated by an angle that depends on both the delay k and the frequency index m.

However, there's no guarantee that an arbitrary filter H will adhere to this structure. Depending on the phase structure of H, the resulting signal could sound completely distorted. Analyzing the phase structure of H can reveal the presence of these distortions.

10.4.1 Examples

Before going into analyzing filter phase, it will help to have some intuitive grasp of how important phase is to a signal.

The examples below demonstrate filters which do not affect the magnitude spectrum at all: $|Y[m]| = |X[m]|$, but the output signal $y = h * x$ is substantially different from the input signal x.

Example 10.1 (Discarding phase). To demonstrate the importance of phase, we can listen to what happens when we discard all phase information in a signal, so that

$$Y[m] = |X[m]|$$

This is equivalent to applying a filter with unit amplitude and phase exactly opposite of $X[m]$:

$$H[m] = \frac{\overline{X[m]}}{|X[m]|}.$$

This can be implemented in Python as follows:

```python
# Take the DFT
X = np.fft.rfft(x)

# Discard phase
Y = np.abs(X)

# Invert the DFT
y = np.fft.irfft(Y)
```

and the results are demonstrated in Fig. 10.9.

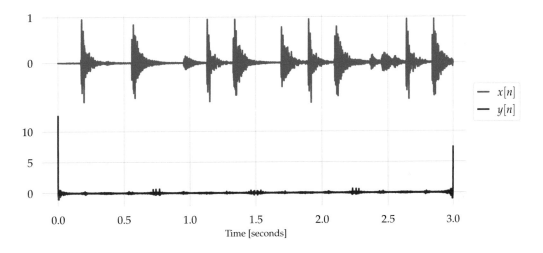

Figure 10.9 A signal $x[n]$ (`https://freesound.org/s/50712/`, top) consisting of a few seconds of a drum loop is reconstructed as $y[n]$ (bottom) by having its phase information discarded. Note that all timing information has vanished in $y[n]$.

Discarding phase information preserves some aspects of the signal: the same frequencies are generally present, but the temporal coherence of the signal has been destroyed.

Example 10.2 (Random phase). The example above is particularly bad, but not unique. As a second example, we can imagine creating a filter H with *random phases* between 0 and 2π.

```
# get the length and DFT of a signal x
N = len(x)
X = np.fft.rfft(x)

# random(N) makes N random numbers between 0 and 1
# multiplying by 2pi gives us random phases
phase = np.random.random(len(X)) * 2 * np.pi
H = np.exp(1j * phase)

# Rotate each X[m] by a random amount
Y = X * H

# recover the time-domain signal
y = np.fft.irfft(Y)
```

The result of this transformation is demonstrated below in Fig. 10.10. In this case, randomizing the phase has rendered the original signal completely unintelligible. **Phase is important.**

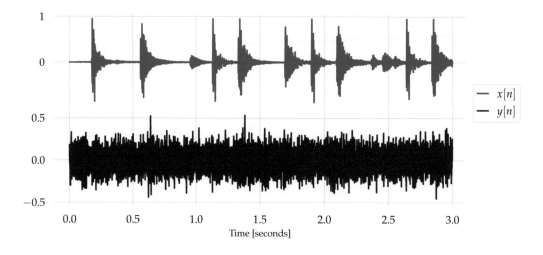

Figure 10.10 A signal $x[n]$ is reconstructed as $y[n]$ (bottom) by having its phase information replaced by random angles. Just as in Fig. 10.9, all timing information has vanished in $y[n]$.

10.4.2 The phase spectrum

Let H denote the DFT spectrum of an impulse response h. Rather than plot the magnitude spectrum $|H|$, the **phase spectrum** is the sequence of phases $\phi[m]$, where

$$H[m] = A[m] \cdot e^{j\phi[m]}.$$

In Python, the phase spectrum can be obtained as follows:

```python
# Take the DFT of the filter
H = np.fft.fft(h)

# Extract the angle: between -pi and +pi
phase = np.angle(H)
```

Plotting $\phi[m]$ as a sequence can help us understand the delay behavior of the filter for each analysis frequency m. However, correctly interpreting the phase spectrum requires a bit more care than interpreting the magnitude spectrum.

If we interpret delay as a convolutional filter, we can plot the phase spectrum of different delays to see what happens. Some examples are displayed below in Fig. 10.11.

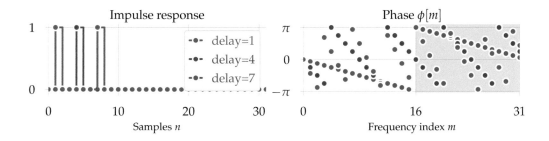

Figure 10.11 **Left**: three different delay filters (1, 4, and 7 samples) are plotted with length $N = 64$. **Right**: for each filter, the phase spectrum $\phi[m]$ is plotted as a function of frequency.

From Fig. 10.11, we can observe that the different delay values make zig-zag shapes with different slopes. This is exactly what is predicted by the DFT shifting theorem, which implies

$$\phi[m] = -2\pi \cdot \frac{k}{N} \cdot m \qquad (10.2)$$

for a k-sample delay. When plotted as a function of m, $\phi[m]$ should produce a line with slope $-2\pi \cdot k/N$, but this is not exactly what we see.

The zig-zag effect is due to angles wrapping around at $\pm\pi$. The np.angle function extracts the angle from a complex number, but there are ambiguities in how angle is interpreted. Remember that $\theta \equiv \theta + 2\pi$ for any angle θ, so np.angle will always take the equivalent angle that lies within the range $[-\pi, +\pi)$.

This effect can be undone by using the np.unwrap function. Angle unwrapping takes a sequence of angles $\phi[m]$ and produces a sequence of equivalent angles $\phi'[m]$ that avoid large discontinuities (greater than π radians) between $\phi[m]$ and $\phi[m + 1]$ by adding multiples of 2π. The unwrapped angles will land outside the $\pm\pi$ range, but satisfy $\phi'[m] = \phi[m] + c \cdot 2\pi$ (for some integer c).

In code, this is done as follows:

```
# Extract the angle: between -pi and +pi
phase = np.angle(H)

# Unwrap the phase
phase_unwrap = np.unwrap(phase)
```

The results of phase unwrapping on the delay filters in Fig. 10.11 are illustrated below.

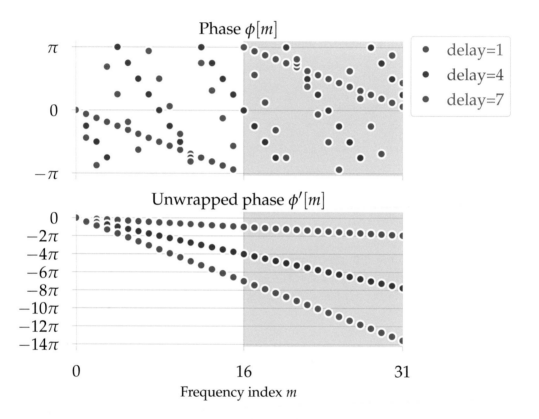

Figure 10.12 **Top**: the phase spectra $\phi[m]$ of three different delay filters (1, 4, and 7 samples) appear as *zig-zag* lines, bounded between $-\pi$ and $+\pi$. **Bottom**: the unwrapped phase spectra $\phi'[m]$ appear as straight lines with slope depending on the amount of delay.

10.4.3 Group delay

The delay examples above illustrate a general principle. If a filter's (unwrapped) phase spectrum is **linear** – i.e., $\phi'[m] = a \cdot m$ for some constant a – then the filter implements a delay that preserves the relative positioning of each sinusoidal component. In other words, filters with linear phase *preserve the phase coherence* of any input signal.

Conversely, a filter with *non-linear* phase, like the *examples above*, will delay different sinusoidal components by different amounts, and can significantly distort the signal. For this reason, linear phase is generally a desirable property.

For delay filters, the delay parameter can be recovered from the phase spectrum by rearranging (10.2) to solve for k:

$$k = -\frac{N}{2\pi \cdot m} \cdot \phi[m], \tag{10.3}$$

and this value would be the same for all frequency indices $m > 0$. The sample delay k can be converted into delay time t (measured in seconds) by dividing by the sampling rate:

$$t = \frac{k}{f_s} = -\frac{N}{2\pi \cdot m \cdot f_s} \cdot \phi[m]. \tag{10.4}$$

Note that the delay t is proportional to the *negative* slope of the phase spectrum.

Dimensional analysis

In reading (10.3), it is helpful to remember the units applied to each quantity:

- $\phi[m]$ [radians]

- m [cycles / signal-duration]

- N [samples / signal-duration]

- 2π [radians / cycle]

So the result k will have units of [samples]. This will be helpful below.

More generally, if someone hands you an arbitrary filter h, you wouldn't know in advance whether it has linear phase or not, and it would be incorrect to apply (10.3) because it is built on the assumption of linear phase. However, we can generalize the idea to relax this assumption.

Instead of assuming linear phase – that $\phi[m] \propto m$, or equivalently, that the slope of the phase response is constant – we can estimate the negative slope separately for each frequency $m > 0$. If the filter truly has linear phase, then all estimates should agree, and give us the desired delay parameter. If the filter does not have linear phase, we can detect this by seeing disagreements in the slope estimates.

The simplest way to estimate slope is the "rise-over-run" method, that is, dividing the change in neighboring (unwrapped) phase measurements (the *rise*)

$$\Delta_\phi[m] = \phi'[m] - \phi'[m-1]$$

by the change in the corresponding frequencies (the *run*, in Hz):

$$\Delta_f[m] = f_s \cdot \frac{m}{N} - f_s \cdot \frac{m-1}{N} = \frac{f_s}{N}.$$

The result is a sequence of negative slope measurements:

$$-\frac{\Delta_\phi[m]}{\Delta_f[m]} = \frac{N}{f_s} \cdot (\phi'[m] - \phi'[m-1]). \tag{10.5}$$

Applying dimensional analysis to (10.5), the numerator $\Delta_\phi[m]$ is a difference of angles measured in [radians], while the denominator $\Delta_f[m]$ is a difference of frequencies measured in [cycles/second]. The ratio therefore has units [radian-second/cycles], which is not entirely easy to understand. However, if we divide by 2π [radians/cycle], the result will have units of [seconds], which nicely corresponds to our intuitive notion of delay.

Putting this all together gives us the formal definition of **group delay**.

Definition 10.2 (Group delay). Let h be an impulse response of length N samples, and let $\phi'[m]$ denote its unwrapped phase spectrum.

The **group delay** of h is the given by the sequence:

$$G[m] = -\frac{N}{2\pi \cdot f_s} \cdot (\phi'[m] - \phi'[m-1]). \tag{10.6}$$

and is measured in [seconds].

For FIR filters, group delay can be calculated in Python as follows:

```
# Take the DFT
H = np.fft.fft(h)

# Extract angles
phase = np.angle(H)

# Unwrap the angles
phase_unwrap = np.unwrap(phase)

# Calculate group delay from the differences in
# successive phase measurements
# We tack the last phase onto to the beginning so that the m=0 case
# is handled correctly, and compares to m=-1.
G = - len(h) / (2 * np.pi * fs) * np.diff(phase_unwrap,␣
↪prepend=phase_unwrap[-1])
```

Alternatively, the `scipy` module provides a function for computing group delay for both FIR and infinite impulse response (IIR) filters, which we will see in the next chapter:

```
# First get our frequencies
frequencies = np.fft.fftfreq(len(h), 1/fs)

# Compute group delay (in samples) using scipy
```

(continues on next page)

(continued from previous page)

```
#      - (h, 1) is how we indicate that this is an FIR filter
#      - w=frequencies is how we specify DFT analysis frequencies
frequencies, G_samples = scipy.signal.group_delay((h, 1),␣
 ↪w=frequencies, fs=fs)

# Convert delay measurements from samples to seconds
G = G_samples / fs
```

10.4.4 Linear and non-linear phase

As mentioned above, not all filters will have completely linear phase. To demonstrate this, let's create two filters, one low-pass and one high-pass, using the window method:

```
fs = 22050
f_cutoff = 3000

# 4 cycles of our cutoff frequency
order = 4 * fs // f_cutoff

h_low = scipy.signal.firwin(order,
                            f_cutoff,
                            window='hann',
                            pass_zero='lowpass',
                            fs=fs)

h_high = scipy.signal.firwin(order,
                             f_cutoff,
                             window='hann',
                             pass_zero='highpass',
                             fs=fs)
```

Fig. 10.13 illustrates the impulse response, frequency response, and unwrapped phase spectrum of these two filters.

Unlike the delay filters depicted in Fig. 10.12, these filters do not have a strictly linear response. In both cases, the unwrapped phase $\phi'[m]$ has a sharp bend near the cutoff frequency f_c, where a strictly linear phase response would continue in the same direction indicated in the pass-band.

However, if we restrict attention to the *pass band* of the filters, they both have linear phase response, as indicated by the solid lines in Fig. 10.13. The non-linear part of the phase response is limited to the stop band, **and this is fine**. Frequencies in the stop band should be attenuated (and ideally, inaudible), and even if phase distortion happens there, it won't matter because we won't hear it.

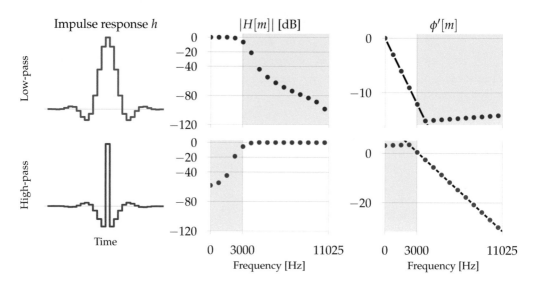

Figure 10.13 Two FIR filters are shown (**left column**), along with their frequency response $|H[m]|$ (**center column**) and unwrapped phase spectrum $\phi'[m]$ (**right column**). Shaded regions correspond to the stop-bands of the filters in the frequency domain. **Top row**: a low-pass filter with $f_c = 3000$ Hz. **Bottom row**: a high-pass filter with $f_c = 3000$ Hz. Neither filter has completely linear phase response, but both are linear when restricted to the pass band, as indicated by the solid line.

For this reason, it is typically sufficient to have linear phase only in the pass band of a filter, and it is common to refer to such filters as having linear phase, even if their entire phase response is technically non-linear.

10.4.5 Summary

In this section, we've seen how to analyze the phase spectrum of FIR filters. This lead to the definition of *group delay*, which allows us to determine A) if a filter is non-linear, and B) how much delay a filter will apply to a signal.

Practically speaking, the filters we've covered in this chapter (window method and Parks-McClellan) are designed to have linear phase in their pass-bands, so subsequent analysis is not strictly necessary to determine linearity. That said, the method demonstrated above can still be useful if you need to infer the delay of a given filter.

In the next chapter, we will introduce *infinite* impulse response filters, which generally will not have linear phase. Understanding group delay in that setting will be much more important, so it's a good idea to get some practice with the method in the FIR setting first.

10.5 EXERCISES

Exercise 10.1. If you have a signal x of length N, and a filter h of length $K < N$, the standard *full-mode* convolution $h*x$ can be implemented by using circular convolution on zero-padded versions of x and h, and trimming the results.

What is the minimal amount of padding (in terms of N and K) that must be applied to x and h for this to work?

Exercise 10.2. Starting from the example code in *Filter Design and Analysis*, experiment with different filter designs to improve the quality of low-pass filters with a cutoff of $f_c = 500$ Hz. In particular:

- For window design filters, what combination of length (order) and window works best? (Try the following windows: `hann`, `hamming`, `rect`, `blackmanharris`)

- For Parks-McClellan (Remez) filters, what combination of length and transition region width works best?

- Base your comparisons on both plots of the frequency response curves, as well as listening to the results on a signal of your choice.

Exercise 10.3. For each of the filters you tested in the previous question, compute the group delay $G[m]$. Each filter will have linear phase in the pass band, but some may continue to have linear phase through the transition band, and possibly into the stop band! For each filter, determine the frequency at which the phase response becomes non-linear, and the corresponding frequency response (in dB) of the filter at that frequency, and collect your results in a table.

- Which filter(s) retain linear phase at the highest frequency?

- What about at the lowest response (in dB)?

- What can this tell you about each filter's perceptual quality?

Exercise 10.4. Prove the dual convolution theorem *Corollary 10.1*. Can you explain where the extra factor of $1/N$ comes from?

## CHAPTER 11

Infinite impulse response filters

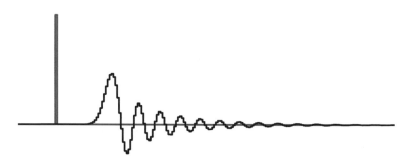

In the previous chapter, we saw how to use convolution to implement finite impulse response (FIR) filters, such as low-pass filters. FIR filters have many nice properties:

- They're easy to implement (by convolution);

- They're stable;

- They can be analyzed by the DFT.

However, one big drawback to FIR filters is that they can be somewhat inefficient. As we saw in *Filter Design and Analysis*, achieving a desired stop-band attenuation can require a filter with thousands of coefficients!

In this chapter, we'll develop *infinite* impulse response filters. This family of filters uses **feedback** to process signals, resulting in more powerful and efficient filters. This comes at the cost of sacrificing the nice properties of FIR filters mentioned above, but don't worry: we'll see in the final chapter how to recover similar properties.

11.1   FEEDBACK FILTERS

The systems we've studied so far (e.g., convolutional filters) are **feed-forward** systems: each output $y[n]$ depends only on the input signal x (and the parameters of the filter). A **feedback** system is one where each output $y[n]$ can also depend on **previous outputs**. The situation is analogous to the feedback encountered by placing a microphone (or, perhaps, an electric guitar) close to its speaker, creating a closed loop that results in resonance, like the example below.

Feedback systems can be powerful, but need to be treated with care. If you place a microphone *too close* to an amplifier's output, the system can become unstable and damage its components, as well as the eardrums of anyone who happens to be nearby.

11.1.1   Example: exponential moving average

As an introduction to feedback filters, let's imagine the following process for computing a moving average of a signal $x[n]$, resulting in output signal $y[n]$:

$$y[n] = \frac{1}{2} \cdot x[n] + \frac{1}{2} \cdot y[n-1]. \tag{11.1}$$

This is a **feedback** process because the output $y[n]$ depends on both the input $x[n]$ (a feed-forward process) and the **previous output** $y[n-1]$ (feedback). Here, we'll take the convention that $y[-1] = 0$, so the first output $y[0] = \frac{1}{2} \cdot x[0]$. An example of this process is illustrated below in Fig. 11.1.

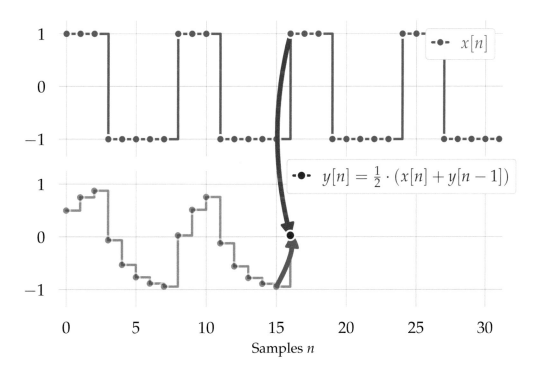

Figure 11.1 **Top**: an input signal $x[n]$. **Bottom**: the output $y[n]$ of an exponential moving average filter (11.1) combines the current input sample $x[n]$ with the previous output sample $y[n-1]$ (*arrows*).

Just like we did with *convolutional filters*, we can try to understand the behavior of (11.1) by providing an impulse signal as input $x = [1, 0, 0, \dots]$ and seeing how the system responds:

$$\begin{aligned}
x[0] &= 1 & \Rightarrow y[0] &= 1/2 \\
x[1] &= 0 & \Rightarrow y[1] &= 1/2 \cdot y[0] = 1/4 \\
x[2] &= 0 & \Rightarrow y[2] &= 1/2 \cdot y[1] = 1/8 \\
x[3] &= 0 & \Rightarrow y[3] &= 1/2 \cdot y[2] = 1/16
\end{aligned}$$

$$\dots$$

$$y[n] = 2^{-(n+1)}$$

This shows that the impulse response, illustrated below in Fig. 11.2, decays *exponentially* (hence the name: *exponential moving average*). However, there is no finite length N for which all $n \geq N$ satisfy $y[n] = 0$: the impulse response is **infinitely long**. This is where the term **infinite impulse response (IIR)** filter comes from.

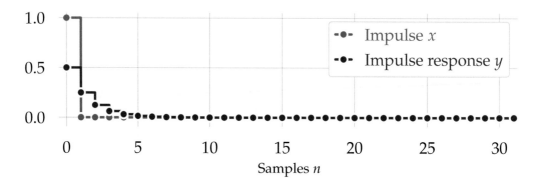

Figure 11.2 The impulse response of (11.1) decays exponentially: $y[n] = 2^{-(n+1)}$, but never reaches 0.

11.1.2 IIR Filters

The example above is one of the simplest non-trivial feedback systems we could write down, but its behavior is far from simple.

Generalizing the idea of incorporating previous outputs to compute the next output leads to the following standard definition of an IIR filter.

Definition 11.1 (Linear IIR filter). A **linear IIR filter** consists of two sets of parameters: the **feed-forward** parameters $b = [b_0, b_1, \dots, b_{K-1}]$, and the **feedback** parameters $a = [a_0, a_1, \dots, a_{K-1}]$.

Given an input signal x, the output is computed according to the following equation:

$$a[0] \cdot y[n] = \sum_{k=0}^{K-1} b[k] \cdot x[n-k] - \sum_{k=1}^{K-1} a[k] \cdot y[n-k]. \tag{11.2}$$

Feed-forward and feed-back?

At this point, you might be wondering why the feed-back parameters are called a and the feed-forward parameters are called b. One would think that b is for **back**, right?

Alas, this notation is traditional, and standard across nearly all signal processing frameworks.

One way to remember this convention is:

- b for **before** the filter is applied,

- a for **after** the filter is applied.

There's much to unpack in this definition, but it's not as scary as it may seem. Let's look at a couple of special cases before moving on.

11.1.3 Special case 1: $a[0] = 1$

Note in (11.2) that the feed-back summation starts at $k = 1$, and the first feedback coefficient $a[0]$ appears on the left-hand side of the equation. This is an admittedly annoying bit of notation, but it is so pervasively used that we'll stick with it. The idea here is that if the summation started at $k = 0$, then $y[n]$ would depend on itself: $y[n - 0] = y[n]$. Keeping the $k = 0$ feedback term separate lets us define $y[n]$ without any such circular dependencies.

As long as $a[0] \neq 0$, we can divide through on both sides to get an equation that isolates the output $y[n]$ in terms of the inputs:

$$y[n] = \frac{1}{a[0]} \cdot \left(\sum_{k=0}^{K-1} b[k] \cdot x[n - k] - \sum_{k=1}^{K-1} a[k] \cdot y[n - k] \right).$$

In code, this could be implemented as follows to compute the first N output samples:

```
def filter(b, a, x):
    '''Apply a linear IIR system defined by b and a to a signal x.

    b and a must have the same length, and a[0] cannot be 0.
    '''
    # Allocate N output samples
    N = len(x)
    y = np.zeros(N)

    K = len(b)  # Get the order of the filter

    # Compute each output y[n] in sequence
    for n in range(N):
        for k in range(min(K, n)):  # Add the feed-forward part
            y[n] += b[k] * x[n-k]
```

(continues on next page)

(continued from previous page)

```
        for k in range(1, min(K, n)):   # Subtract the feed-back part
            y[n] -= a[k] * y[n-k]
        y[n] /= a[0]   # Divide through by a[0]
    return y
```

When $a[0] = 1$, this leading factor of $1/a[0]$ disappears, and we're left with the equation

$$y[n] = \sum_{k=0}^{K-1} b[k] \cdot x[n-k] - \sum_{k=1}^{K-1} a[k] \cdot y[n-k].$$

This case is particularly common, because we can always replace the coefficients $b[k]$ by $b[k]/a[0]$ (and $a[k]$ by $a[k]/a[0]$) without changing the behavior of the system. In many implementations, it therefore always assumed that $a[0] = 1$.

Returning to the first example, we can implement (11.1) with the following coefficients:

$$b = \begin{bmatrix} 1 \\ \frac{1}{2} \end{bmatrix}$$

$$a = \begin{bmatrix} 1, -\frac{1}{2} \end{bmatrix}.$$

11.1.4   Special case 2: $a[k] = 0$

If all of the feedback coefficients $a[k > 0] = 0$, and $a[0] = 1$ as above, then the IIR definition (11.2) simplifies to a *standard convolution*:

$$y[n] = \sum_{k=0}^{K-1} b[k] \cdot x[n-k].$$

FIR filters, therefore, are a special case of IIR filters. Everything we develop with IIR filters can also be applied to FIR filters.

11.1.5   Causal filters

All of the IIR filters that we'll encounter define the output $y[n]$ in terms of inputs up to $x[n]$ and previous outputs up to (and including) $y[n-1]$. Note that $y[n]$ does not depend on $x[n+1]$, or on $y[n+1]$.

Systems with this property are known as **causal systems**. It is also possible to have *non-causal* systems, where the output at sample index n can depend on future inputs like $x[n+1]$. This might sound strange, but we'll see later in this chapter that there are sometimes good reasons to use non-causal filters.

11.1.6   Why are feedbacks subtracted instead of added?

Another apparently strange quirk of IIR notation is that the feed-forward sum is *added*, but the feed-back sum is *subtracted*. At first glance, this seems arbitrary and weird, and often leads to mistakes in implementation.

The true reason for this convention will become apparent in the next chapter, when we cover the *z-transform*. For now, it may be helpful to think about what would happen if we tried to isolate all of the y terms on one side of the equation, and the x terms on the other:

$$\sum_{k=0}^{K-1} a[k] \cdot y[n-k] = \sum_{k=0}^{K-1} b[k] \cdot x[n-k]. \tag{11.3}$$

The key benefit of subtracting the feedback terms, rather than adding them, is that when we isolate y from x as in (11.3), all terms appear *positively*. Although this form cannot be used directly to compute each output $y[n]$, it will help prevent sign errors and generally make our lives easier later on when we develop the z-transform.

11.1.7 Summary

The use of feedback can be a conceptual leap for many people, so let's now pause and take stock of what we've seen so far.

- Feedback systems can have an infinite impulse response, even when the input is finite in length.

- We can always assume the first feedback coefficient $a[0] = 1$.

- IIR filters generalize FIR filters.

- The feed-back terms are subtracted, not added, to produce each output $y[n]$.

- In general, b and a need not be the same length, but we can always pad the shorter of the two with zeros to make them match.

With these facts in mind, we can now proceed to trying out some examples.

11.2 USING IIR FILTERS

In the previous section, we saw the general definition of IIR filters. In this section, we'll see how to use them in practice, and begin to probe at understanding their behavior by testing their outputs in response to synthetic input signals.

As in the previous chapter, we'll focus on low-pass filters, but the ideas will readily generalize to other settings.

11.2.1 IIR filters in Python

In the previous section, we defined a function *filter* that implements (11.2). Given an input signal x and filter coefficients b (feed-forward) and a (feedback), we construct the output signal y one sample at a time, using nested loops in a manner not too different from (time-domain) convolution.

In practice, we would do better to use an existing implementation and not write our own from scratch. Luckily, the `scipy.signal` package provides exactly this function: `lfilter` (linear filter).

For example, the exponential moving average filter could be implemented as follows:

```
import scipy.signal

b = [1/2]   # Feed-forward
a = [1, -1/2]   # Feed-back

# Apply the filter to input signal x
y = scipy.signal.lfilter(b, a, x)
```

We can use this, for example, to compute the first few samples of the impulse response:

```
# Compute the first few samples of the impulse response
# of an exponential moving average

x = np.array([1, 0, 0, 0, 0, 0, 0])
y = scipy.signal.lfilter([1/2], [1, -1/2], x)
print(y)
```

```
[0.5        0.25      0.125     0.0625    0.03125   0.015625  0.
 ↪0078125]
```

11.2.2 Example: Butterworth filters

The first type of filter that we'll look at is called the *Butterworth filter*, after Stephen Butterworth [B+30]. We'll not get into the details of how the filter coefficients are defined, but instead rely on the `scipy.signal.butter` function to construct them for us.

Using `scipy.signal.butter` is not too different from using the window-method function `firwin` to design an FIR low-pass filter. We'll need to supply the *order* of the filter (the number of coefficients), as well as the cutoff frequency f_c and the sampling rate f_s. The example code below constructs an order-10 filter with $f_c = 500$ Hz and $f_S = 44100$.

```
fs = 44100   # Sampling rate
fc = 500   # Cutoff frequency

# Build the low-pass filter
b, a = scipy.signal.butter(10, fc, fs=fs)

# Print the coefficients
print('Feed-forward: ', b)
print('Feed-back:    ', a)
```

```
Feed-forward:  [2.62867578e-15 2.62867578e-14 1.18290410e-13 3.
↪15441093e-13
 5.52021913e-13 6.62426296e-13 5.52021913e-13 3.15441093e-13
 1.18290410e-13 2.62867578e-14 2.62867578e-15]
Feed-back:     [  1.            -9.54462136    41.00487909 -104.
↪41737082   174.53480697
 -200.09486144  159.34094444  -87.02835847   31.2004603    -6.6300023
    0.6341236 ]
```

To demonstrate the filter's behavior, let's apply it to a 250-sample delayed impulse, padded with zeros out to 1000 samples. There's nothing special about the particular delay that we're using, but it will allow us to compare the magnitude and phase spectra before and after applying the filter. Remember that impulses (including delayed impulses) have energy at all frequencies, so this should give us a sense of how well the filter works at attenuating high frequencies. By default, the `lfilter` function produces an output y of the same length as the input, which might not capture all of the response behavior that we want to look at. Padding the input with trailing zeros gives us time to observe more of the response without truncating prematurely.

Warning. In this section, we're investigating the behavior of an IIR filter by analyzing a finite bit of its output in response to a particular input signal. This should not be confused with analyzing the filter directly. We'll see how to do that in the next chapter.

```
# Create the delayed impulse signal
x = np.zeros(1000)
x[249] = 1

# Apply the Butterworth filter from above
y = scipy.signal.lfilter(b, a, x)
```

We can now inspect the DFT of the input x and output y to see what the filter has done: the results are illustrated in Fig. 11.3.

We can immediately observe a couple of interesting things in Fig. 11.3, especially when compared to the FIR filters demonstrated in Figs. 10.7 and 10.8. First, the time-domain output y is delayed relative to the input–it peaks much later–with asymmetric ripples that decay to 0.

Second, even though the filter has only order 10, its stop-band attenuation is comparable to the FIR filters, which have much higher order (in the hundreds).

Third, the phase response looks *approximately* linear in the pass-band, but it turns out not to be *exactly* linear. As a result, different frequencies present in x have been delayed by slightly different amounts to create y, which we perhaps could have anticipated from the fact that x is symmetric in time (after padding) but y is asymmetric.

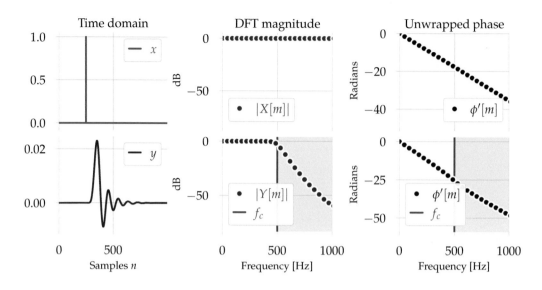

Figure 11.3 **Top-left**: a 250-sample delay signal x, padded with zeros to 1000 samples. **Top-center**: the magnitude spectrum $|X|$, measured in decibels, zoomed into the frequency range $[0, 1000]$. **Top-right**: the unwrapped phase spectrum ϕ' of X. **Bottom-left**: the output y after applying an order-10 low-pass Butterworth filter with cutoff $f_c = 500$. **Bottom-center**: the magnitude spectrum $|Y|$, with stop-band marked by the shaded region. **Bottom-right**: the unwrapped phase spectrum ϕ' of Y.

These observations point to general features of IIR filters (not just Butterworth): they can be much more efficient than FIR filters, but they rarely have linear phase. Applying them naively can lead to phase distortion artifacts.

11.2.3   Selecting the order

In the example above, the choice of order 10 might seem arbitrary, but it was in fact chosen to satisfy certain criteria: no more than 3dB pass-band ripple, 60 dB of stop-band attenuation, and a transition band from 500 to 1000 Hz. The `scipy` helper function `scipy.signal.buttord` takes these parameters, and provides the minimal filter order which satisfies the constraints:

```
fc = 500   # Cutoff at 500 Hz
fstop = 1000   # End the transition band at 1000 Hz
passband_ripple = 3   # 3dB ripple
stopband_atten = 60   # 60 dB attenuation of the stop-band

# Get the smallest order for the filter
order = scipy.signal.buttord(fc, fstop, passband_ripple, stopband_
 ↪atten, fs=fs)
```

(continues on next page)

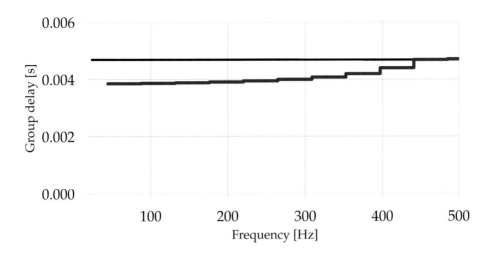

Figure 11.4 A more careful look at the group delay of y within the passband shows that the delay is not constant over all frequencies, especially near the cutoff $f_c = 500$.

(continued from previous page)

```
# Now construct the filter
b, a = scipy.signal.butter(order, fc, fs=fs)
```

If you recall how *The Parks-McClellan method* optimizes the filter coefficients subject to constraints on ripple, attenuation, and transition bandwidth, the idea is similar here (although the underlying optimization is quite different). In general, when using IIR filters, it's a good idea to look for helper functions (like order selection) which can help you determine the best settings to satisfy your constraints.

11.2.4 Compensating for phase distortion

The example above illustrates two things: 1) that IIR filters induce delay, just like FIR filters, and 2) IIR filters are often not linear-phase, resulting in phase distortion (see margin Fig. 11.4).

There is a nice trick for dealing with both of these issues: filter the signal twice, but in *opposite directions*. More specifically, we do the following:

1. Apply the filter once: `y1 = filter(b, a, x)`

2. Let `y1_rev` be `y1` in reverse order: `y1_rev = y1[::-1]`

3. Apply the filter again: `y2_rev = filter(b, a, y1_rev)`

4. Reverse again to obtain the final output: `y = y2_rev[::-1]`

The key idea here is that if filtering the signal once (forward in time) induces some delay, then filtering it again in the opposite direction (backward in time) should undo that delay exactly. This even works if the delay is different for each frequency, so even if a filter has non-linear phase, we can undo any distortion artifacts.

There are two caveats to be aware of.

1. Any gain that would be applied by the filter will be applied *twice*. If a frequency is attenuated by 40 dB in one pass, it will be attenuated by 80 dB in two passes. This also goes for gain and pass-band ripple, not just stop-band attenuation.

2. Double-filtering can only be applied if we can observe the entire input signal in advance: it is **non-causal**. This is fine for pre-recorded signals, but cannot work for real-time audio streams.

Caveats aside, this technique is so common that it is also provided by most signal processing frameworks. In `scipy`, it is called `scipy.signal.filtfilt`, and we can use it just like we would use `lfilter`. The example below illustrates how to apply this method to our example signal and low-pass filter, with results illustrated in Fig. 11.5.

```
fs = 44100   # Sampling rate
fc = 500   # Cutoff frequency

# Build the low-pass filter for double-filtering
# This means we'll cut the pass-band ripple and stop-band␣
␣attenuation both in half
# We only need the first output from buttord
order, _ = scipy.signal.buttord(fc, 1000, 3/2, 60/2, fs=fs)

b2, a2 = scipy.signal.butter(order, fc, fs=fs)

# Apply the double filter
y2 = scipy.signal.filtfilt(b2, a2, x)
```

Fig. 11.5 illustrates that the double-filtering approach does effectively align the final output y_2 to the input x: their time-domain representations peak in exactly the same position. Note, however, that reverse-filtering does introduce *pre-echo*: the oscillations in y_2 preceding the peak at $n = 249$. In this respect, the result is similar to the FIR filter outputs in Fig. 10.8, and is to some extent unavoidable if we're applying low-pass filtering to signals with strong transients. Remember: although the outputs are similar, we got there with much less work: an order-6 IIR filter applied twice, compares well to FIR filters with order in the hundreds.

Butterworth filters are just the beginning: in the next section, we'll meet a few more types of IIR filters.

> **Tip.** If you plan to use double-filtering, design your filters to have half the target gain. That is, if you want your total stop-band attenuation to be 80dB, design your filter to achieve 40dB.

11.3   COMMON IIR FILTERS

In the last section, we were introduced to the *Butterworth* filter, which is one of the earliest and most commonly used IIR filters. In this section, we'll meet a few more

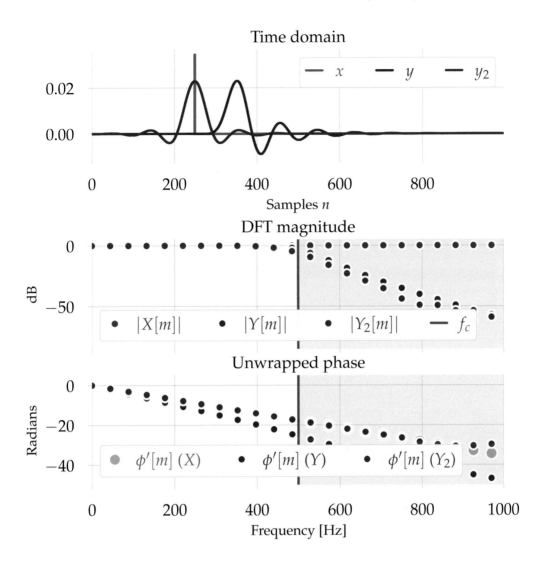

Figure 11.5 **Top**: the input signal x (an impulse delayed by 250 samples), along with the single-pass output y (order=10) and the double-pass output y_2 (order=6). **Middle**: the DFT magnitudes of X (constant), Y, and Y_2. **Bottom**: the unwrapped phase of X, Y, and Y_2. Note that the phase of Y_2 agrees with X over the pass-band, indicating a total delay of 0.

types of IIR filter. None of them are necessarily better or worse than the others: they each have benefits and drawbacks. Having a sense of the various trade-offs made by each type of filter is important if you plan to apply these in practice.

11.3.1   Chebyshev filters

The next family of filters that we'll see are known as *Chebyshev* filters, named for the mathematician Pafnuty Chebyshev, who initially developed the underlying family of polynomial functions upon which the filters are based. There are two types of Chebyshev filters, not-so-conveniently named *Type 1* and *Type 2*.

Type 1

Type 1 Chebyshev filters have a steeper transition than Butterworth filters, at the expense of introducing ripple in the pass-band. They are constructed by the function `scipy.signal.cheby1`, which has as its key parameters:

- the order of the filter, which can be derived by `scipy.signal.cheby1ord`,

- the maximum amount of passband ripple we'll allow, and

- the cutoff frequency.

Continuing our example from the previous section, a 500 Hz low-pass filter could be constructed as follows:

```
fs = 44100   # Sampling rate
fc = 500   # Cutoff frequency
fstop = 1000   # stop-band at 1000

ripple = 3   # we'll allow 3 dB ripple in the passband
attenuation = 60   # we'll require 60 dB attenuation in the stop band

# Get the order and discard the second output (natural frequency)
order, _ = scipy.signal.cheb1ord(fc, fstop, ripple, attenuation,␣
↪fs=fs)

# Build the filter
b, a = scipy.signal.cheby1(order, ripple, fc, fs=fs)
```

Fig. 11.6 illustrates the results of applying this filter to an impulse x, and how it compares to a Butterworth filter.

As Fig. 11.6 illustrates, the frequency response drops much more rapidly after f_c than that of the Butterworth filter. The cost of this is twofold:

1. There is ripple in the pass-band: $|Y|$ is not constant for frequencies below f_c.

2. There is more phase distortion: ϕ' deviates significantly from linear phase (plotted as a straight line).

Type 2

Type 2 Chebyshev filters are similar to type 1, except the pass-band remains flat and ripple is allowed in the *stop-band*. Type 2 filters are constructed by `scipy.signal.cheby2`, which is similar to the `cheby1` function above, but with two key differences:

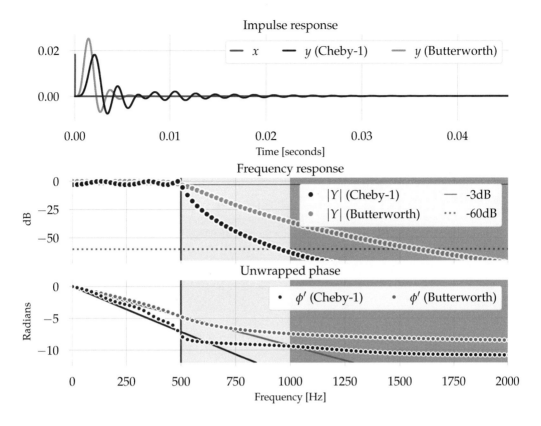

Figure 11.6 A type 1 Chebyshev filter with 3dB passband ripple, cutoff frequency of $f_c = 500$, and stop-band starting at 1000 Hz is applied to an impulse x to produce the impulse response y, measured for 2000 samples (at $f_s = 44100$). For comparison purposes, a Butterworth filter with the same cutoff and order is also computed. **Top**: the impulse, and impulse response. **Middle**: the DFT magnitude $|Y|$ measured in decibels. **Bottom**: the unwrapped phase spectrum ϕ'.

1. The stop-band attenuation is provided, instead of the pass-band ripple, and

2. The **end** of the transition region is provided, rather than the cutoff frequency f_c.

Similarly, the order of the filter is computed by `scipy.signal.cheb2ord`, which takes all the same parameters as `cheb1ord`.

```
# Get the order and discard the second output (natural frequency)
order2, _ = scipy.signal.cheb2ord(fc, fstop, ripple, attenuation,␣
↪fs=fs)

# Build the filter
b2, a2 = scipy.signal.cheby2(order2, attenuation, fstop, fs=fs)
```

Fig. 11.7 illustrates the behavior of the type 2 filter, compared to the type 1 filter with similar constraints.

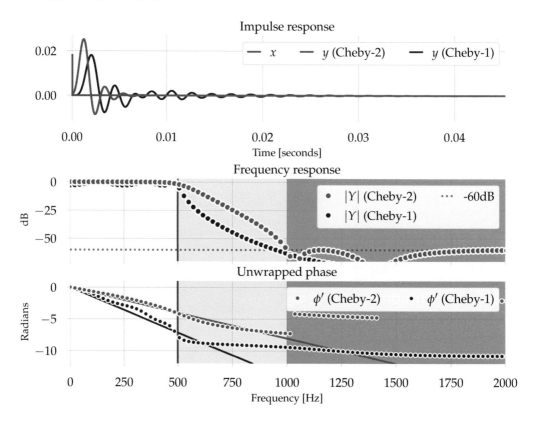

Figure 11.7 A type 2 Chebyshev filter with 60dB stop-band attenuation, cutoff frequency of $f_c = 500$, and stop-band starting at 1000 Hz. The type 1 filter from Fig. 11.6 is included for comparison purposes. **Top**: the impulse, and impulse response. **Middle**: the DFT magnitude $|Y|$ measured in decibels. **Bottom**: the unwrapped phase spectrum ϕ'.

Compared to the type 1 filter, Fig. 11.7 shows that the type 2 filter has a shallower transition, though it is still steeper than that of the Butterworth filter. Note that the pass-band frequency response has no ripple, and the phase is much closer to linear. We should therefore expect better preservation of the low-frequency content with the type 2 filter than with the type 1 filter. The main cost is that the transition region has less attenuation, so the filter will not be as *sharp*, and frequencies above f_c will propagate through slightly more with the type 2 filter.

11.3.2 Elliptic filters

So far, we've seen:

- the Butterworth filter, which has no ripple in either the pass-band or the stop-band;

- the type 1 Chebyshev filter, which has pass-band ripple but no stop-band ripple; and

- the type 2 Chebyshev filter, which has stop-band ripple but no pass-band ripple.

If we allow ripple in *both* the pass- and stop-bands, we get what are known as **elliptic** filters. The main benefit of elliptic filters is that they can have extremely steep transitions.

Elliptic filters are provided by `scipy.signal.ellip` (and with order helper `scipy.signal.ellipord`). The `ellip` function can be thought of as generalizing both `cheby1` and `cheby2`, so it requires all of the parameters of either individual function:

- the order of the filter;

- the pass-band ripple;

- the stop-band attenuation; and

- the cutoff frequency (like in `cheby1`, not the end of the transition like in `cheby2`!)

The `ellipord` helper function works just like `cheby1` and `cheby2`, and the following code can be used to construct an elliptic low-pass filter:

```
# Get the order and discard the second output (natural frequency)
order_ell, _  = scipy.signal.ellipord(fc, fstop, ripple, attenuation,
↪ fs=fs)

# Build the filter
b_ell, a_ell = scipy.signal.ellip(order_ell, ripple, attenuation, fc,
↪ fs=fs)
```

Fig. 11.8 shows that allowing ripple in both the pass- and stop-bands allows the Elliptic filter to achieve a much steeper transition than either type 1 or type 2 Chebyshev filters. Note, however, that this also results in quite a bit of phase distortion in the Elliptic filter.

11.3.3 Summary

The examples above are not exhaustive, and merely illustrate a few features of each of the filters in question. Notably, we did not vary the ripple, attenuation, or transition region constraints here, and these parameters have enormous influence on the behavior of the filters. For example, allowing more ripple (type 1 or elliptic) can provide a steeper transition.

Compared to FIR filters, IIR filters can be more efficient, achieving comparable performance (e.g., stop-band attenuation) at a fraction of the computational cost dictated by the order. This makes IIR filters an attractive choice for real-time processing applications, where computational efficiency and latency are critical. However,

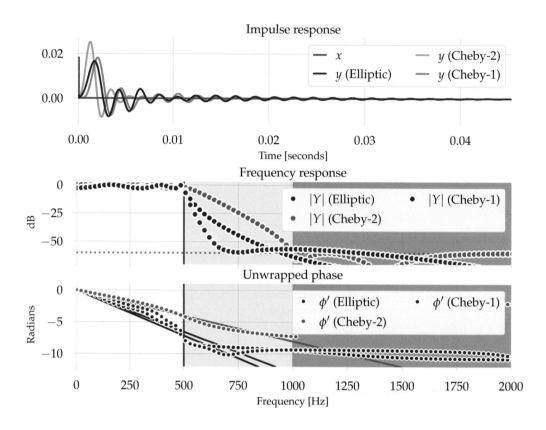

Figure 11.8 An elliptic filter with 3dB pass-band ripple, 60dB stop-band attenuation, cutoff frequency of $f_c = 500$, and stop-band starting at 1000 Hz. Chebyshev filters are included for comparison purposes. **Top**: the impulse, and impulse response. **Middle**: the DFT magnitude $|Y|$ measured in decibels. **Bottom**: the unwrapped phase spectrum ϕ'.

we've also seen that IIR filters can have non-linear phase response, which may not be desirable depending on the application.

So which filter should you use? **It depends!** (I know, not so helpful of an answer.) A few things to keep in mind:

1. If you do not want any attenuation of pass-band frequencies, stick to Butterworth or type 2 Chebyshev.

2. How much phase distortion can you tolerate? Or can you compensate for it by using bidirectional filtering?

3. If you need steep transitions, Elliptic might be the best option.

The table below summarizes the properties of the four filter types we've seen.

Filter type	Pass-band ripple	Stop-band ripple	Transition	Phase distortion
Butterworth	✗	✗	Slow	Small
Chebyshev (Type 1)	✓	✗	Medium	Medium
Chebyshev (Type 2)	✗	✓	Medium	Small
Elliptic	✓	✓	Fast	Large

11.4 EXERCISES

Exercise 11.1. Consider the following IIR system:

$$y[n] = \frac{1}{2}(x[n] - x[n-2]) + y[n-1]$$

1. What is the order of this system?

2. Put this system into the standard form of (11.2). What are b and a for this system?

3. What is its impulse response?

Exercise 11.2. Use `scipy.signal.butter` to construct a **high-pass** filter at (and $f_s = 44100$) with the following properties:

- $f_c = 500$

- a transition width of 250 Hz,

- at most 1 dB of pass-band loss, and

- stop-band attenuation of 60 dB.

What is the order of the resulting filter?

Exercise 11.3. You are asked to implement a low-pass filter at $f_s = 44,100$ with the following specification:

- Cut-off frequency $f_c = 1000$ Hz

- No more than 1.5 dB of ripple in the pass-band (less is acceptable)

- At least 80 dB of attenuation in the stop-band (more is acceptable)

- A transition band of no more than 500 Hz (less is acceptable)

1. Which of the IIR filter(s) covered in this chapter could you use to implement a filter with these constraints?

2. Using the helper functions provided by scipy (e.g., `scipy.signal.ellipord`), determine the order of the filter. If there are multiple filter types that will work, which one gives the smallest order?

3. Use the *Parks-McClellan method* to implement an FIR filter with the given constraints. What order does the filter need to be to satisfy the constraints?

For this question, stick to single-pass (`lfilter`) filtering, and not bidirectional (`filtfilt`) filtering.

Hint. For part (c), the `remez` function does not allow explicit constraints on the attenuation. Try starting with a small order, measuring the frequency response, and then increasing the order until the constraints are satisfied.

## CHAPTER 12

Analyzing IIR filters

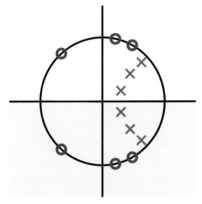

The previous chapter introduced infinite impulse response (IIR) filters, which are defined recursively as feedback loops. These filters can be both more powerful and more efficient than FIR (convolutional) filters. However, these benefits come with a cost: IIR filters cannot be analyzed directly by the discrete Fourier transform.

In this chapter, we'll develop a new analytical tool, the *z-transform*, which will allow us to analyze IIR systems.

12.1 THE Z-TRANSFORM

The infinite impulse response filters that we've encountered in the previous chapter can be thought of as a generalization of convolutional, finite impulse response filters. We first saw convolution in the time domain, where it has the interpretation of mixing delayed and gained copies of the input signal. We then saw how in the frequency domain, this is equivalent to taking the element-wise product of the Fourier transforms of the signal and filter coefficients. This observation provided the tools we need to understand the effect that any given convolutional filter will have when applied to an input signal.

This leads us to the central question of this chapter: **can we analyze IIR filters in the frequency domain as well?**

It turns out that the answer is **yes**, but *not with the discrete Fourier transform*. We'll need a new tool, which is known as the **z-transform**!

DOI: 10.1201/9781003264859-12

12.1.1 Why can't we use the DFT?

It's worth taking a step back, and reconstructing the chain of reasoning that led us to the discrete Fourier transform and the convolution theorem.

The DFT is defined in terms of **analysis frequencies**: those which complete an integral number of cycles over a fixed duration of time (N samples). The DFT is then constructed by comparing the similarity (sum-product) of the input signal $x[n]$ to a collection of complex exponentials at the chosen analysis frequencies. Remember: the DFT always has the same number of analysis frequencies N as input samples.

The convolution theorem provides a way to understand convolution $y = h * x$ in terms of the per-frequency DFT products $X[m] \cdot H[m]$. For this to work, the two DFT sequences $X[m]$ and $H[m]$ must have the same (finite) length N. This is not a significant barrier: both x and h must be finite length for the DFT to be defined in the first place, and we can always zero-pad the shorter signal to match the longer signal.

However, when we have an IIR filter, this does not work. Even for relatively simple filters, like the exponential moving average

$$y[n] = 1/2 \cdot x[n] + 1/2 \cdot y[n-1],$$

the impulse response can be infinite in length: there is no finite N for which all $y[n > N] = 0$. As a result, there is no finite N from which we can determine the analysis frequencies, which we would need to construct the DFT!

12.1.2 Defining the z-transform

The limitations of the DFT ultimately all come from the dependence on knowing the length of the signal N. Therefore, to remove these limitations, we will need to modify the DFT definition everywhere the length N appears.

Recall the definition of the DFT for frequency index m:

$$X[m] = \sum_{n=0}^{N-1} x[n] \cdot e^{-\mathrm{j} \cdot 2\pi \cdot \frac{m}{N} \cdot n}.$$

The length N appears in two places here: once in the limit of the summation ($N-1$), and once in the complex exponential.

The first step we will take is to rewrite the DFT definition by using the exponent multiplication rule

$$e^{a \cdot b} = (e^a)^b.$$

Specifically, we can rewrite the complex exponential by factoring out $-n$ as a power:

$$e^{-\mathrm{j} \cdot 2\pi \cdot \frac{m}{N} \cdot n} = \left(e^{2\pi \cdot \mathrm{j} \cdot \frac{m}{N}}\right)^{-n}.$$

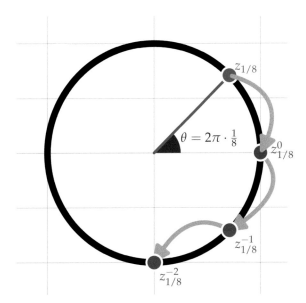

Figure 12.1 When the complex number $z_{1/8} = \exp(\mathrm{j} \cdot 2\pi \cdot 1/8)$ is raised to successive powers $z_{1/8}^{-n}$ for $n = 0, 1, 2, \ldots$, the resulting effect is to move the point clockwise around the unit circle by steps of $2\pi \cdot 1/8$ radians. This general pattern holds for any $z_{m/N}$ with integers m and N, which provide the basis for the discrete Fourier transform.

We can now interpret this as a sequence (parametrized by sample index n) generated by a complex number $z_{m/N}$ on the unit circle at angle $\theta = 2\pi \cdot m/N$, or

$$z_{m/N} = e^{2\pi \cdot \mathrm{j} \cdot \frac{m}{N}}. \tag{12.1}$$

The ratio m/N can thus be interpreted as the fraction of the circle we travel in each step n of the sequence, as shown in Figure 12.1: when m is a small integer, the steps are small, so it will take more steps to traverse the entire circle. That is, small m corresponds to low frequency, as we should expect. The length N determines the smallest non-zero step $(1/N)$, and as N increases, we can take smaller steps.

If we take (12.1), we can rewrite the DFT calculation as a polynomial in $z_{m/N}$:

$$X[m] = \sum_{n=0}^{N-1} x[n] \cdot (z_{m/N})^{-n}.$$

If we now define $x[n < 0] = x[n \geq N] = 0$, so that the input signal is assumed to be silent outside the observed set of samples, we can relax the finite summation for $n = 0 \ldots N - 1$ to an infinite summation:

$$X[m] = \sum_{n=0}^{\infty} x[n] \cdot (z_{m/N})^{-n}.$$

This form is still equivalent to the DFT, but under the assumption of silence after N samples, rather than infinite repetition.

The final step is to replace $z_{m/N}$ (which correspond to our analysis frequencies for $m = 0, 1, 2, \ldots, N - 1$) by *any* complex number z. In doing so, we have completely eliminated the dependence on the signal duration N. The resulting transformation of $x[n]$ is no longer a sequence of N complex numbers $X[0], X[1], \ldots, X[N - 1]$, but instead defines a continuous function $X(z)$ which can be evaluated for *any* complex number z, hence the name: **z-transform**.

Definition 12.1 (The z-transform). Given a sequence $x[n]$ (of possibly infinite length), the z-transform is a function $X(z)$ over complex numbers z defined as follows:

$$X(z) = \sum_{n=0}^{\infty} x[n] \cdot z^{-n}. \tag{12.2}$$

12.1.3 Relating the z-transform to the DFT

The first thing to be absolutely clear about is that the z-transform does not "replace" the DFT exactly. In particular, we don't typically use the z-transform to analyze a signal $x[n]$ directly. As we'll see shortly, the z-transform is more useful for analyzing filters, and this works essentially by comparing the z-transforms of the input $X(z)$ and the output $Y(z)$.

Unlike the DFT, the z-transform does not produce a finite sequence of coefficients. Instead, it produces a *function* that can be evaluated for any complex number z. This function $X(z)$ – note the use of parentheses (z) instead of square brackets $[n]$ – is not really an object you would use directly, and it is primarily a conceptual device that we use for analysis. For this reason, you are unlikely to encounter software implementations that compute (12.2) for you.

However, the DFT can be recovered from the z-transform by evaluating $X(z)$ at carefully chosen points z. Specifically, the DFT coefficients $X[m]$ can be extracted from the z-transform $X(z)$ as follows:

$$X[m] = X\left(e^{2\pi \cdot \mathrm{j} \cdot m/N}\right).$$

12.1.4 Aside: continuous frequency and the DTFT

Another special case of the z-transform is known as the *discrete-time Fourier transform* (DTFT). The DTFT is similar to our conventional DFT, except that instead of using N analysis frequencies $0, f_s/N, 2 \cdot f_s/N, \cdots$, it is defined for **any** frequency f. The DTFT gets its name from the fact that it is continuous in frequency, but still discrete in time.

The DTFT for any frequency f is obtained by evaluating the z-transform at a point on the unit circle with angle $2\pi \cdot f/f_s$:

$$\mathrm{DTFT}(x)(f) = X\left(e^{\mathrm{j} \cdot 2\pi \cdot f/f_s}\right).$$

For our purposes, the DTFT is primarily useful as a way to use the z-transform for arbitrary frequencies, not just analysis frequencies. Note that the DTFT also makes the connection between frequency and angle precise: a frequency f with sampling rate f_s is associated with a complex number of unit magnitude and angle $2\pi \cdot f/f_s$. For example:

- frequency $f = 0$ is associated with angle 0;

- aliasing frequencies map to the same angle, since

$$2\pi \cdot \frac{f + k \cdot f_s}{f_s} = 2\pi \cdot \frac{f}{f_s} + 2\pi \cdot k \equiv 2\pi \cdot \frac{f}{f_s},$$

- and the Nyquist frequency $f = f_s/2$ is associated with angle $2\pi \cdot (f_s/2)/f_s = \pi \equiv -\pi$.

This generalizes the DFT's analysis frequencies, in that using $f = m \cdot f_s/N$ exactly recovers the points $z_{m/N}$ defined above in (12.1). More generally, the range of frequencies $0 \leq f \leq f_s$ are mapped continuously to the range of angles $0 \leq \theta \leq 2\pi$ as shown in Figure 12.2. This will become useful later on, when we use the z-transform to understand frequency response of feedback filters.

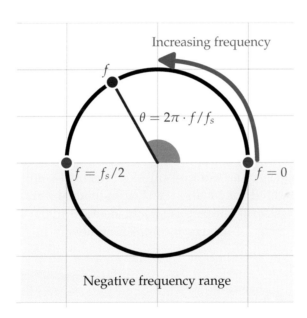

Figure 12.2   The z-transform (and discrete-time Fourier transform) associates each continuous frequency f with a specific angle $\theta = 2\pi \cdot f/f_s$. Angles in the upper half of the complex plane correspond to positive frequencies between 0 and $f_s/2$ (the Nyquist frequency). Angles in the lower half correspond to negative frequencies between 0 and $-f_s/2 \equiv f_s/2$.

12.1.5   Summary

We've now introduced the fundamental tool that we'll need to analyze IIR filters. At this point, you might be wondering how or why this idea is useful. On the one hand, we've generalized the DFT in a way that no longer depends on the signal length. On the other hand, the resulting transformation doesn't actually produce a finite representation of the signal, and is not something that we could explicitly compute anyway, since it encodes a function $X(z)$ on the entire complex plane $\mathbb{C}$.

As we will see in the coming sections, the power of the z-transform is that it allows us to reason algebraically about signals and filters – even infinitely long signals – in terms of polynomials (e.g., $z^2 - z + 1$). In the next section, we'll establish the properties of the z-transform that make this work.

12.2   PROPERTIES OF THE Z-TRANSFORM

Like the discrete Fourier transform, the z-transform has many convenient mathematical properties. In this section, we'll establish the most important ones: linearity, shifting, and convolution. These properties should all be conceptually familiar from our earlier study of the DFT; however, they may look a little different in our new context.

12.2.1   Linearity

Theorem 12.1 (The z-transform is linear). Let $y[n] = c_1 \cdot x_1[n] + c_2 \cdot x_2[n]$ denote a combination of two signals x_1 and x_2, each with gain coefficients $c_1, c_2 \in \mathbb{R}$.

Then the z-transform of $y[n]$ is equivalent to

$$Y(z) = c_1 \cdot X_1(z) + c_2 \cdot X_2(z),$$

where $X_1(z)$ and $X_2(z)$ are the z-transforms of $x_1[n]$ and $x_2[n]$ respectively.

Or, in plain language: the z-transform of a weighted combination of signals is equivalent to the weighted combination of their individual z-transforms.

The proof of z-transform linearity is almost identical to our earlier proof of DFT *Linearity*, but now using an arbitrary complex number z in place of the complex sinusoid of the DFT.

Proof.

$$Y(z) = \sum_{n=0}^{\infty} y[n] \cdot z^{-n} \qquad\qquad \text{definition of z-transform}$$

$$= \sum_{n=0}^{\infty} (c_1 \cdot x_1[n] + c_2 \cdot x_2[n]) \cdot z^{-n} \qquad\qquad \text{definition of } y[n]$$

$$= \sum_{n=0}^{\infty} c_1 \cdot x_1[n] \cdot z^{-n} + c_2 \cdot x_2[n] \cdot z^{-n} \qquad\qquad \text{distributing } z^{-n} \text{ over sum}$$

$$= c_1 \cdot \left(\sum_{n=0}^{\infty} x_1[n] \cdot z^{-n} \right) + c_2 \cdot \left(\sum_{n=0}^{\infty} x_2[n] \cdot z^{-n} \right) \quad c_1, c_2 \text{ do not depend on } n$$

$$= c_1 \cdot X_1(z) + c_2 \cdot X_2(z) \qquad\qquad \text{definition of z-transform}$$

$\square$

12.2.2 Shifting theorem

The shifting theorem for z-transforms looks a little different from the DFT shifting theorem, but it's conceptually quite similar. The main distinction is that we no longer have the circular/repetition assumption, and instead assume silence outside the observed samples.

Theorem 12.2 (z-transform shifting theorem). Let $y[n] = x[n - k]$ denote a k-step delay of a signal $x[n]$, with the assumption that $x[n < 0] = 0$.

Then the z-transform of $y[n]$ is given by

$$Y(z) = z^{-k} \cdot X(z). \tag{12.3}$$

Proof. Let $y[n] = x[n - k]$. Since we're assuming $x[n < 0] = 0$ (silence before the signal starts), we must have $y[n < k] = 0$.

The z-transform is then given by

$$
\begin{aligned}
Y(z) &= \sum_{n=0}^{\infty} y[n] \cdot z^{-n} && \text{definition of z-transform} \\
&= \sum_{n=k}^{\infty} y[n] \cdot z^{-n} && \text{since } y[n < k] = 0 \\
&= \sum_{n=k}^{\infty} x[n - k] \cdot z^{-n} && \text{definition of } y[n] \\
&= \sum_{n=0}^{\infty} x[n] \cdot z^{-(n+k)}
\end{aligned}
$$

This last step follows from the observation that each z^{-n} pairs with the sample $x[n - k]$, which means that each $x[n]$ pairs with $z^{-(n+k)}$.

Continuing the proof, we can factor

$$
\begin{aligned}
z^{-(n+k)} &= z^{-n-k} \\
&= z^{-n} \cdot z^{-k},
\end{aligned}
$$

so that

$$
\begin{aligned}
Y(z) &= \sum_{n=0}^{\infty} x[n] \cdot z^{-(n+k)} \\
&= \sum_{n=0}^{\infty} x[n] \cdot z^{-n} \cdot z^{-k} \\
&= z^{-k} \cdot \sum_{n=0}^{\infty} x[n] \cdot z^{-n} && z^{-k} \text{ does not depend on } n \\
&= z^{-k} \cdot X(z) && \text{definition of z-transform.}
\end{aligned}
$$

□

Remark 12.1 (The DFT shifting theorem). Remember that the DFT can be derived from the z-transform by letting $z = \exp\left(2\pi \cdot \mathrm{j} \cdot m/N\right)$. If we plug this into the z-transform shifting theorem, then a delay k results in multiplication by z^{-k}, which in this case gives us

$$\left(\exp\left(2\pi \cdot \mathrm{j} \cdot m/N\right)\right)^{-k} = \exp\left(-2\pi \cdot \mathrm{j} \cdot k \cdot m/N\right),$$

exactly as stated by the DFT shifting theorem!

12.2.3   Convolution theorem

Just like the DFT, we also have a convolution theorem for the z-transform, and its proof follows basically the same form as that of the DFT convolution theorem, except relying on the z-transform shifting theorem where appropriate.

Theorem 12.3 (z-transform convolution theorem). Let

$$y[n] = \sum_{k=0}^{K-1} h[k] \cdot x[n-k]$$

denote the convolution of a signal $x[n]$ with an impulse response $h[k]$ of length K. We will assume that $h[k < 0] = h[k \geq K] = 0$.

The z-transform of the convolution y is equal to the product of the z-transforms:

$$Y(z) = H(z) \cdot X(z), \tag{12.4}$$

where $Y(z)$, $H(z)$, and $X(z)$ are the z-transforms of $y[n]$, $h[k]$, and $x[n]$, respectively.

Proof. Let $y[n] = \sum_{k=0}^{K-1} h[k] \cdot x[n-k]$. Taking the z-transform of both sides gives us:

$$Y(z) = \mathrm{ZT}\left(\sum_{k=0}^{K-1} h[k] \cdot x[n-k]\right)$$

$$= \sum_{k=0}^{K-1} h[k] \cdot \mathrm{ZT}(x[n-k]) \qquad \text{by linearity of the z-transform}$$

$$= \sum_{k=0}^{K-1} h[k] \cdot z^{-k} \cdot X(z) \qquad \text{by the z-transform shifting theorem}$$

$$= X(z) \cdot \sum_{k=0}^{K-1} h[k] \cdot z^{-k} \qquad X(z) \text{ does not depend on } k$$

$$= X(z) \cdot \sum_{k=0}^{\infty} h[k] \cdot z^{-k} \qquad h[k \geq K] = 0, \text{ so summands past } k = K \text{ add } 0$$

$$= X(z) \cdot H(z) \qquad \text{by definition of z-transform.}$$

$\square$

12.2.4 Summary

In this section, we've revisited the core properties of the DFT that are necessary for understanding FIR filters: linearity, shifting, and convolution. The derivations for the z-transform versions of these theorems are not substantially different from what we've seen previously with the DFT, though some of the notation and underlying assumptions have changed to fit our new setting.

In the next section, we'll see how to apply these properties in the analysis of IIR filters.

12.3 TRANSFER FUNCTIONS

When we previously analyzed FIR filters, it became useful to examine the discrete Fourier transform $H[m]$ of the impulse response $h[k]$. The frequency domain view of convolutional filters immediately exposes how a filter will delay and gain (or attenuate) different frequencies. However, as noted earlier in this chapter, the DFT cannot be applied directly to feedback systems with infinite impulse responses, and this led to our derivation of the z-transform.

In this section, we'll return to our initial motivation, and derive something similar to $H[m]$, except from the perspective of the z-transform rather than the DFT. The resulting object is known as the *transfer function* of a filter and is denoted by $H(z)$.

12.3.1 Defining a transfer function

Let's start with a linear IIR filter in standard form:

$$a[0] \cdot y[n] = \sum_{k=0}^{K-1} b[k] \cdot x[n-k] - \sum_{k=1}^{K-1} a[k] \cdot y[n-k],$$

where $b[k]$ and $a[k]$ denote the feed-forward and feedback coefficients of the filter.

> This is the point where the benefit of including $a[0]$ and subtracting the feedback terms (rather than adding them) pays off. (12.6) would have been much more awkward without $a[0]$ and the correct sign on the feedback terms!

If we move all feedback terms to the left-hand side of the equation, we obtain an equivalent equation:

$$\sum_{k=0}^{K-1} a[k] \cdot y[n-k] = \sum_{k=0}^{K-1} b[k] \cdot x[n-k]. \tag{12.5}$$

While this form is not useful for *computing* y, it is useful for **analyzing** y!

In particular, you might recognize that both sides of the equation are convolutions:

$$a * y = b * x. \tag{12.6}$$

This means that if we take the z-transform of both sides, we can use the z-transform convolution theorem:

$$A(z) \cdot Y(z) = B(z) \cdot X(z),$$

where $A(z), Y(z), B(z), X(z)$ denote the z-transforms of a, y, b, and x respectively.

As long as $A(z)$ is not zero – and it generally is non-zero except for at most K specific choices of z – we can divide through to isolate $Y(z)$:

$$Y(z) = \frac{B(z)}{A(z)} \cdot X(z).$$

This gives us something highly reminiscent of the convolution theorem: filtering in the time domain has again been expressed as multiplication, except now in the z-plane instead of the frequency domain.

Definition 12.2 (Transfer function). Let a and b denote the feedback and feed-forward coefficients of a linear IIR filter:

$$a[0] \cdot y[n] = \sum_{k=0}^{K-1} b[k] \cdot x[n-k] - \sum_{k=1}^{K-1} a[k] \cdot y[n-k].$$

The **transfer function** of this filter is defined as follows

$$H(z) = \frac{B(z)}{A(z)} = \frac{\displaystyle\sum_{k=0}^{K-1} b[k] \cdot z^{-k}}{\displaystyle\sum_{k=0}^{K-1} a[k] \cdot z^{-k}} \tag{12.7}$$

12.3.2 Using transfer functions

As mentioned previously, transfer functions can be thought of as providing a generalization of the convolution theorem to support feedback filters. Specifically, we have

$$Y(z) = H(z) \cdot X(z) = \frac{B(z)}{A(z)} \cdot X(z). \tag{12.8}$$

Note that as a special case, if $a = [1]$ (so that there is no feedback in the filter), we get the z-transform

$$A(z) = \sum_{k=0}^{\infty} a[k] \cdot z^{-k} = a[0] \cdot z^0 = a[0] = 1.$$

In this case, the transfer function simplifies to $H(z) = B(z)/1 = B(z)$, and we recover the convolution theorem.

More generally, we can still reason about $H(z)$ as the object that transforms the input signal x into the output signal y. Evaluating $H(z)$ at values of z with unit magnitude – i.e. $z = e^{j\cdot\theta}$ – produces the *frequency response* of the filter. In Python, this is provided by the function `scipy.signal.freqz` (frequency response via z-transform).

Example: analyzing a Butterworth filter

In *Using IIR filters*, we constructed a Butterworth filter and rather crudely analyzed its effect on an impulse input by truncating the output y and taking the DFT. Let's revisit this example, but instead, analyze it using the z-transform and transfer functions.

```
fs = 44100    # 44.1K sampling rate
fc = 500      # 500 Hz cutoff frequency
order = 10    # order-10 filter

b, a = scipy.signal.butter(order, fc, fs=fs)

freq, H = scipy.signal.freqz(b, a, fs=fs)
```

The result of this computation is an array `freq` of frequencies (spaced uniformly between 0 and `fs/2`), and an array `H` that evaluates the transfer function at each specified frequency.

How many frequencies?

By default, scipy's `freqz` uses `worN=512` frequencies (up to, but not including Nyquist), but this can be easily changed to whatever frequency resolution you prefer. Often, we are only using `freqz` to support a visual inspection of the frequency response, and for that use case, 512 frequencies is typically sufficient.

Fig. 12.3 illustrates the response curve by plotting `freq` on the horizontal axis and `abs(h)` (in decibel scale) on the vertical axis, i.e.:

```
import matplotlib.pyplot as plt

# Convert |H| to decibels, with a -120dB noise floor
H_dB = 20 * np.log10(np.abs(H) + 1e-6)
plt.plot(freq, H_dB)
```

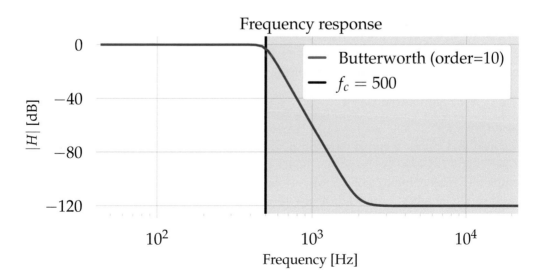

Figure 12.3 The frequency response curve $|H(z)|$ for an order-10 Butterworth filter with a cutoff frequency at $f_c = 500$ Hz and sampling rate $f_s = 44{,}100$.

Note that we never had to construct a test signal to generate this curve: all of the information was inferred from the filter coefficients b and a!

Example: phase response of an elliptic filter

Just like with convolutional filters, we can infer delay properties of IIR filters by looking at the phase spectrum contained in $H(z)$. This can be done directly, e.g., by calling `scipy.signal.freqz` and then using `np.unwrap(np.angle(h))` to compute the unwrapped phase. Alternatively, it is usually more convenient to use the already provided `group_delay` function, as demonstrated below for an elliptic filter, and visualized in Figure 12.4.

```
fs = 44100   # Sampling rate
fc = 500   # Cutoff frequency
fstop = 600
attenuation = 40   # we'll require 40 dB attenuation in the stop band
ripple = 3   # we'll allow 3 dB ripple in the passband

order_ell, wn = scipy.signal.ellipord(fc, fstop, ripple, attenuation,
↪ fs=fs)
b_ell, a_ell = scipy.signal.ellip(order_ell, ripple, attenuation, wn,
↪ fs=fs)

# scipy's group delay measures in samples
# we'll convert to seconds by dividing out the sampling rate
freq, delay = scipy.signal.group_delay([b_ell, a_ell], fs=fs)
delay_sec = delay / fs
```

Note that `scipy.signal.group_delay` returns the delay for each measured frequency in `freq` in units of `[samples]`. To convert the delay measurements to `[seconds]`, we divide by the sampling rate `fs`. A visualization of the resulting group delay (along with the frequency response) is provided below.

12.3.3   Why not use the DTFT?

Everything we've done so far only depends on z with unit magnitude: frequency response and phase response are both properties of $H\left(e^{j\cdot\theta}\right)$, which we can think of as computing the discrete-time Fourier transform (DTFT). At this point, it is completely reasonable to wonder why we needed to define the z-transform to support any complex number z, instead of just those points on the unit circle.

As we will see in the next section, expanding the definition to include all complex z allows us to study the *stability* of a filter, and this would not be possible with just the DTFT.

12.4   STABILITY, POLES, AND ZEROS

Most of the filters we've encountered so far have been *stable*, in the sense that if an input signal $x[n]$ has bounded values, so too will the output $y[n]$.

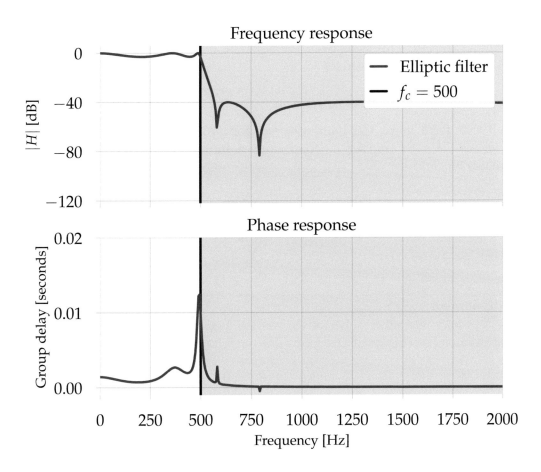

Figure 12.4 The frequency response curve $|H(z)|$ (*top*) and phase response curve (group delay, *bottom*) for an elliptic filter with cutoff $f_c = 500$, transition bandwidth of 100 Hz, 3dB passband ripple, and 40dB stop-band attenuation ($f_s = 44{,}100$). The plot is limited to the frequency range [0, 2000] so that the trends within the passband are easier to see.

As an example of an unstable system, consider

$$y[n] = x[n] + 2 \cdot y[n-1]. \tag{12.9}$$

Plugging $x[n] = [1, 0, 0, 0, \ldots]$ into this definition produces the impulse response

$$
\begin{aligned}
y[0] &= x[0] + 2 \cdot 0 & &= 1 \\
y[1] &= x[1] + 2 \cdot y[0] & &= 2 \\
y[2] &= x[2] + 2 \cdot y[1] & &= 4 \\
&\quad\cdots \\
y[n] &= 2^n
\end{aligned}
$$

which grows exponentially and without bound. This system is **unstable**.

While it is relatively straightforward to determine that (12.9) is unstable, it is not always so easy. This leads us to the question: how can we determine in general whether a given system is stable?

12.4.1 Transfer functions and polynomials

Recall that for a linear IIR filter with feed-forward coefficients $b[k]$ and feedback coefficients $a[k]$, the transfer function is given by the ratio of the z-transform of sequences $b[k]$ and $a[k]$:

$$H(z) = \frac{B(z)}{A(z)} = \frac{\sum_{k=0}^{K-1} b[k] \cdot z^{-k}}{\sum_{k=0}^{K-1} a[k] \cdot z^{-k}}.$$

Polynomials

You may remember polynomials from an algebra class in the distant past. For example, solving quadratic forms like the following

$$2x^2 - x - 1 = 0$$

either by factoring or by using the quadratic formula.

Quadratics are a special case of polynomials where the largest *degree* is 2. The polynomials that come up in signal processing are usually of much larger degree (given by the order of the filter), and we'll generally be using the computer to find the roots automatically.

More specifically, $H(z)$ is the result of dividing two **polynomials** $B(z)$ and $A(z)$:

$$B(z) = b[0] \cdot z^0 + b[1] \cdot z^{-1} + b[2] \cdot z^{-2} + \cdots + b[K-1] \cdot z^{-(K-1)}$$
$$A(z) = a[0] \cdot z^0 + a[1] \cdot z^{-1} + a[2] \cdot z^{-2} + \cdots + a[K-1] \cdot z^{-(K-1)}.$$

Strictly speaking, these are polynomials in z^{-1} (since the powers have negative exponents instead of positive exponents), but the distinction is unimportant for our purposes.

It turns out that we can learn quite a bit about a filter's behavior by examining the properties of these polynomials, and specifically, finding their **roots** by solving $B(z) = 0$ and $A(z) = 0$.

Roots and algebra

From the fundamental theorem of algebra, we have that any polynomial

$$Q(z) = \sum_{d=0}^{D} q[d] \cdot z^d$$

has D (complex) roots, that is, solutions to the equation $Q(z) = 0$.

For example, the polynomial $2x^2 - x - 1$ has roots $x = 1$ and $x = -1/2$. These roots can be found by factoring the polynomial:

$$2x^2 - x - 1 = (2x + 1) \cdot (x - 1)$$

and then setting each factor to 0 independently (since either being 0 is sufficient to ensure the product is 0):

$$2x + 1 = 0 \Rightarrow x = -\frac{1}{2}$$
$$x - 1 = 0 \Rightarrow x = 1$$

As a second example, consider

$$x^2 - 6x + 9 = 0.$$

This also is a degree 2 polynomial, so it should have two solutions. However, if we factor the polynomial, it turns out that the solutions are not unique:

$$x^2 - 6x + 9 = (x - 3) \cdot (x - 3).$$

In this case, we say that there is one solution $x = 3$ *of multiplicity 2* (because it occurs twice).

The fundamental theorem of algebra does not require that all solutions be unique.

As a final example, take

$$x^3 + 1 = 0$$

or equivalently, $x^3 = -1$. In this case, there is only one *real-valued* solution ($x = -1$), but there are two more *complex-valued* solutions: $x = e^{j \cdot \pi/3}$ and $x = e^{-j \cdot \pi/3}$.

The fundamental theorem of algebra allows for roots to be complex.

12.4.2   Zeros

For the moment, let's focus on the numerator of the transfer function, i.e., the polynomial $B(z)$ derived from the feed-forward coefficients. By definition, if z_0 is a root of $B(z)$, then $B(z_0) = 0$. As long as z_0 is not also a root of $A(z)$, we will have

$$H(z_0) = \frac{B(z_0)}{A(z_0)} = \frac{0}{A(z_0)} = 0.$$

The location in the complex plane of the zeros (roots of $B(z)$) tells us about which frequencies are attenuated by the filter. This is because polynomials are *continuous*: if $B(z_0) = 0$, then other values of z near z_0 will be small ($B(z) \approx 0$).

Before going further, it may be helpful to visualize what we're talking about. To make things concrete, we'll analyze a Type 2 Chebyshev filter constructed by the following code block:

```
fs = 8000   # Sampling rate
fc = 1000   # Cutoff frequency
fstop = 1500 # Stop-band 1KHz above cutoff
attenuation = 80   # we'll require 80 dB attenuation in the stop band
ripple = 6   # we'll allow 6 dB ripple in the passband

# Compute the order for this filter, which in this case is 9
order, wn = scipy.signal.cheb2ord(fc, fstop, ripple, attenuation,␣
↪fs=fs)
b, a = scipy.signal.cheby2(order, attenuation, wn, fs=fs)
```

Once we have the filter coefficients b and a, we can obtain the zeros of the filter by the following code:

```
zeros, _, _ = scipy.signal.tf2zpk(b, a)
```

The function `tf2zpk` takes in a transfer function (represented by b and a) and returns the zeros (z), poles (p), and gain (k) of the filter. We'll come back to poles and gain later on, so for now we'll retain just the zeros.

If we print the zeros, we'll see a sequence of 9 complex numbers:

```
array([ 0.38141995-0.92440187j,   0.26658886-0.96381034j,
       -0.02490664-0.99968978j,  -0.57559933-0.81773187j,
       -0.57559933+0.81773187j,  -0.02490664+0.99968978j,
        0.26658886+0.96381034j,   0.38141995+0.92440187j,
       -1.         +0.j          ])
```

We can then plot these values in the complex plane, and compare them to the frequency response of the filter (generated by `freqz` as in the previous section).

From Fig. 12.5, we can observe that the minima in the frequency response curve line up exactly with the zeros of the transfer function. More generally, frequencies *near* the zeros are also attenuated. You can think of the zeros as physical weights that drag down the frequency response.

This is only part of the story though: we'll also need to look at the feedback coefficients to get the full picture.

12.4.3 Poles

So far, we've found that the zeros of $B(z)$ can tell us about the frequency response of a filter. But what about the zeros of $A(z)$?

The roots of $A(z)$ are known as **poles** of the filter, and are denoted by $p_0, p_1, \ldots, p_{K-1} \in \mathbb{C}$. For any pole p (that is not also a root of $B(z)$), $H(p) = B(p)/A(p)$ would divide by zero. Without getting too far into the technical details, it is possible to make sense of this division by looking at the behavior of $H(z)$ in a neighborhood around a pole.

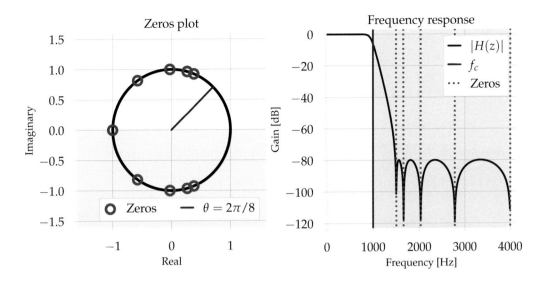

Figure 12.5 **Left**: The zeros of an order-9 Type 2 Chebyshev filter with cutoff frequency $f_c = 1000$, sampling rate $f_s = 8000$, transition bandwidth of 500 Hz, passband ripple of 6dB, and stop-band attenuation of 80dB. The angle $\theta = 2\pi/8$ corresponds to the cutoff frequency divided by the sampling rate $f_c/f_s = 1000/8000$. All zeros of this filter have unit magnitude, and come in conjugate pairs: 4 with positive imaginary component, 4 with negative, and one with zero (at Nyquist). **Right**: The frequency response curve of this filter $|H(z)|$. Each zero in the left plot corresponds to a particular frequency, which coincide with the minima of the frequency response curve as indicated by the dashed lines.

The intuition behind the name "pole" derives from the idea of the function $|H(z)|$ (the magnitude of $H(z)$) can be thought of as a sheet draped over the complex plane, which is held up by (tent) poles. A simplified illustration of this is provided below in Fig. 12.6.

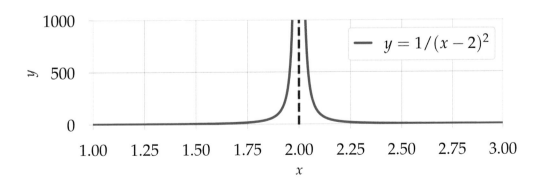

Figure 12.6 The function $y = \frac{1}{(x-2)^2}$ has a pole at $x = 2$. As x approaches 2 from either direction, y increases.

If the zeros of $H(z)$ tell us where frequencies are *attenuated*, the poles tell us where frequencies are **amplified**. (This is a bit of a simplification, as both poles and zeros can affect both gain and attenuation, but let's go with it for now.)

Continuing our previous example, we'll use the `tf2zpk` function to compute the zeros *and* poles for our filter:

```
zeros, poles, gain = scipy.signal.tf2zpk(b, a)
```

Again, this will produce an array of 9 complex numbers for `poles`:

```
array([0.6640386 -0.61973512j,  0.54573487-0.49475044j,
       0.44310112-0.35506889j,  0.36783278-0.18847207j,
       0.3394446 +0.j         ,  0.36783278+0.18847207j,
       0.44310112+0.35506889j,  0.54573487+0.49475044j,
       0.6640386 +0.61973512j])
```

corresponding to the 9 roots of $A(z)$. Like we did above with the zeros plot, we can visualize the position of the poles in the complex plane. Typically, both the poles and zeros are illustrated in the same figure, which is helpfully known as a **pole-zero plot**.

> **Plotting poles and zeros**
>
> Traditionally, poles are denoted by × (crosses), and zeros are denoted by ○ (circles).

Once you learn how to read pole-zero plots, they can be a great way to quickly understand the behavior of a system. For example, in Fig. 12.7, we have zeros at high frequencies (angles approaching π) and poles near low frequencies, so we can infer that this is a low-pass filter. A high-pass filter would exhibit the reverse behavior, with poles at larger angles and zeros at smaller angles.

The fact that zeros land exactly on the unit circle tells us that some frequencies will be attenuated severely (practically to zero).

> **Gain**
>
> The feedback coefficient $a[0]$ is known as the *gain* of the system, as it is used to scale up (or down) the output $y[n]$. As we have noted before, there is some ambiguity in IIR filter design, as we can always divide the remaining coefficients by $a[0]$ without changing the behavior of the system.
>
> This is why defining just the zeros and poles of a system isn't quite enough to fully determine the coefficients. To reconcile this ambiguity, most implementations for zero and pole calculation (e.g., `tf2zpk`) will return the gain ($a[0]$) as a separate return value.

12.4.4   Factored transfer functions

Once we have the zeros z_k and poles p_k of a system, these can be used to write down the transfer function $H(z)$ equivalently in a *factored* form:

$$H(z) = \frac{(z - z_0) \cdot (z - z_1) \ldots (z - z_{K-1})}{(z - p_0) \cdot (z - p_1) \ldots (z - p_{K-1})}$$

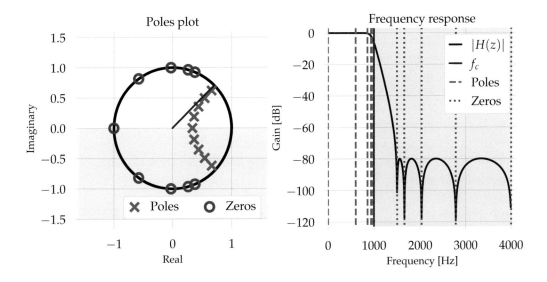

Figure 12.7 **Left**: The poles (×) and zeros (∘) of the filter previously analyzed in Fig. 12.5. **Right**: the frequency response curve, now with both pole and zero angles (frequencies) marked. Note that the poles of this system are not on the unit circle, which is why the frequency response curve does not spike upward to infinity at each pole.

While this doesn't change the definition of the filter – it is just another way of writing the same transfer function – this form does have some benefits.

First, it allows us to reason about high-order filters as a cascade of low-order filters:

$$H(z) = \left(\frac{z - z_0}{z - p_0}\right) \cdot \left(\frac{z - z_1}{z - p_1}\right) \cdots \left(\frac{z - z_{K-1}}{z - p_{K-1}}\right)$$
$$= H_0(z) \cdot H_1(z) \ldots H_{K-1}(z).$$

This observation is often used to simplify the implementation of IIR filters and improve numerical stability (without changing the filter's behavior). Most commonly, this is done by the *second-order sections* (SOS) method, e.g.:

```
# Construct a filter using second-order sections instead of
# feed-forward / feed-back coefficients b and a
sos = scipy.signal.cheby2(order, attenuation, wn, fs=fs, output='sos
↪')

# Instead of lfilter with b and a, use sosfilt
y = scipy.signal.sosfilt(sos, x)

# Or for bidirectional filtering:
y = scipy.signal.sosfiltfilt(sos, x)
```

> **Canceling poles and zeros**
>   Technically, if a point is both a pole and a zero, this produces a factor of $0/0$ in the transfer function. $0/0$ is usually a sign that something bad is happening. However, in this specific case, it does work out.

Second, it provides a way to reason about the interactions between poles and zeros. Notably, if we have a pole which is also a zero, that is, $p_k = z_\ell$, then both of their corresponding factors can be removed from the filter without changing its behavior. Concretely, this means that if a pole and a zero coincide, then their effects cancel each other out.

12.4.5   Stability

Let's return to our first example of an unstable system, $y[n] = x[n] + 2 \cdot y[n-1]$. This is *unstable* because the output can diverge when given an input signal with finite values, e.g., a unit impulse produces a sequence $y[n] = 2^n$ that grows without bound and never settles down.

In standard form, this system has coefficients

$$b = [1]$$
$$a = [1, -2].$$

The transfer function of this system is

$$H(z) = \frac{b[0] \cdot z^0}{a[0] \cdot z^0 + a[1] \cdot z^{-1}} = \frac{1}{1 - 2 \cdot z^{-1}}$$

Because the numerator has degree 0, this system has no zeros. It has one pole, which we can find by solving $1 - 2z^{-1} = 0$:

$$1 - 2z^{-1} = 0 \quad \Rightarrow \frac{2}{z} = 1 \quad \Rightarrow z = 2.$$

Contrast this filter with the first IIR filter we encountered in the previous chapter, the *exponential moving average* (11.1):

$$y[n] = \frac{1}{2} \cdot x[n] + \frac{1}{2} \cdot y[n-1],$$

which has coefficients $b = [1/2]$ and $a = [1, -1/2]$. This filter also has no zeros, and one pole, but the pole is located at $z = 1/2$. It turns out that the exponential moving average filter is *stable*: its output does not diverge for any input signal $x[n]$ with finite values.

So what's different about these two examples?

It turns out that the stability of a filter depends on the location of the poles, and specifically, on their *magnitude*.

Note (IIR stability). Let $H(z)$ denote the transfer function of a linear, IIR filter with zeros $z_0, z_1, \ldots, z_{K-1}$ and poles $p_0, p_1, \ldots, p_{K-1}$, and assume that no pole is also a zero.

The filter is **stable** if **all** poles p have magnitude $|p| < 1$. Or, equivalently, if all poles lie within the unit circle in the complex plane.

The filter is **unstable** if **any** pole p has magnitude $|p| > 1$.

This fact gives us a simple test for stability of a filter:

1. Compute the poles of the filter, e.g., by `scipy.signal.tf2zpk` or `scipy.signal.sos2zpk`.

2. Discard any poles which are also zeros.

3. Compute the magnitude of all remaining poles.

4. If any pole has magnitude larger than 1, the filter is unstable.

As a corollary to this observation, note that FIR filters have no poles (since $A(z) = a[0] = 1 \neq 0$). Since they have no poles, they also have no poles with magnitude larger than 1. This is how we can say that any FIR filter is stable.

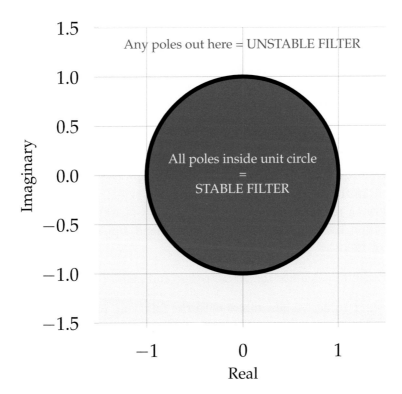

Figure 12.8 If all poles are contained within the unit circle, that is $|p| < 1$ whenever $A(p) = 0$, then the system is stable.

12.4.6 Summary

In this section, we've seen how the z-transform allows us to reason about the behavior of a filter through its transfer function. Specifically, we examined the polynomials

$B(z)$ and $A(z)$ which make up the transfer function $H(z) = B(z)/A(z)$. The roots of these polynomials determine when the transfer function goes to 0 (when $B(z) = 0$, the *zeros*) and when it diverges to infinity ($A(z) = 0$, the *poles*). Finally, the location of the poles of a filter (inside or outside the unit circle) determines whether the filter is stable or unstable.

12.5   EXERCISES

Exercise 12.1. Compute the poles and zeros of each of the following filters. Which ones are stable?

1. $y[n] = \frac{1}{2} \cdot x[n] + \frac{1}{2} \cdot y[n-1] - y[n-2]$

2. $y[n] = x[n] + x[n-2] - 2 \cdot y[n-1]$

Exercise 12.2. A system is given to you with poles at $\{0.5 + 0.5j, 0.5 - 0.5j\}$ and zeros at $\{-0.5, +0.5\}$.

- Create a pole-zero plot for this system, similar to Fig. 12.7 (left).

- Is the system stable?

- Is the system high-pass, low-pass, or neither?

Exercise 12.3. We've seen the function `scipy.signal.tf2zpk`, which produces the zeros, poles, and gain of a filter when given the feed-forward and feed-back coefficients b and a. There is also a function which reverses this process, `scipy.signal.zpk2tf`.

Starting from the example code given in Section 12.3.2, construct an elliptic filter with the following properties: sampling rate of $f_s = 44{,}100$, cutoff frequency of $f_c = 16{,}000$, transition bandwidth of 2000 Hz, pass-band ripple of 3dB, and stop-band attenuation of 120dB.

- Use `tf2zpk` to calculate the zeros, poles, and gain of the filter.

- Use `zpk2tf` to derive new filters with the following properties:

    - `poles = []` (i.e., no poles), but with zeros and gain from the original filter;

    - `zeros = []` (i.e., no zeros), but with poles and gain from the original filter.

    - Substitute the poles for zeros and vice versa, e.g., `scipy.signal.zpk2tf` `(poles, zeros, gain)`.

For each of these three filters (original, no-poles, no-zeros, swapped), plot the resulting frequency response curves given by `scipy.signal.freqz`. How do the modified filters differ from the original?

Mathematical fundamentals

This section of the appendix establishes some mathematical notation and may serve as a brief refresher on fundamental concepts:

- Sets and number systems

- Sequences and summations

- Modular arithmetic

- Exponents and logarithms

A.1 SETS

A *set* is a basic concept in mathematics used to define collections of *elements*. The technical underpinnings of set theory can get a bit tricky, but for our purposes, you can think of a set as any un-ordered collection.

Notationally, we use curly braces {} to denote a set. For example, we could have a set consisting of three colors defined as

$$S = \{\text{red, green, blue}\}.$$

Each of the three colors **red, green, and blue** are *elements* of the set S. The *order* of the elements in a set does not matter: the set above is equivalent to the set {blue, green, red}. *Repetition* also does not matter: an element is either in the set or not.

We use the symbol $\in$ (a funny-looking E, denoting *element*) to denote membership in a set:

$$\text{red} \in S,$$

DOI: 10.1201/9781003264859-A

and $\notin$ to denote that an element does not belong to a set:

$$\text{Cleveland} \notin S.$$

We won't do too much with sets in this text, but the basic notation is helpful to have, especially when dealing with different types of numbers.

A.1.1 Number systems

In digital signal processing, we use many kinds of numbers to represent different quantities. It's helpful to have notation to specify exactly what kind of numbers we're talking about, so here's a brief list with their standard notations:

How natural is zero?

Note: some authors do not include 0 in the natural numbers $\mathbb{N}$. You are likely to encounter differences of opinion when reading other sources, so beware! I've adopted the convention of $0 \in \mathbb{N}$ here for many reasons, but principally because it simplifies notation overall and aligns closely to how we use these numbers when programming.

- $\mathbb{N}$, the *natural numbers*: $\{0, 1, 2, 3, \dots\}$

- $\mathbb{Z}$, the *integers*: $\{\dots, -3, -2, -1, 0, 1, 2, 3, \dots\}$

- $\mathbb{Q}$, the *rational numbers* (fractions): $\{\frac{n}{m} \mid n, m \in \mathbb{Z}, m \neq 0\}$

- $\mathbb{R}$, the *real numbers* (i.e., the continuous number line)

- $\mathbb{C}$, the *complex numbers*: $\{a + jb \mid a, b \in \mathbb{R}\}$

where $j = \sqrt{-1}$ is the *imaginary unit*.

Natural numbers ($\mathbb{N}$) are often used for whole number quantities, such as sample positions n. Note that this means that the first sample will occur at index $n = 0$!

Real numbers ($\mathbb{R}$) are often used for continuous quantities, such as angles (in radians), frequencies (in cycles/sec), or time (in seconds).

Complex numbers occupy a special place in signal processing because they turn out to be a great tool for modeling periodic processes.

A.2 SEQUENCES

A *sequence* is exactly what it sounds like: an ordered list of things. Sequences are different from *sets* in that they are ordered, and repetition is allowed.

The most common usage of sequences in signal processing is the sequence of sample values representing a digital signal. This is commonly denoted by $x[n]$, where $n \in \mathbb{N}$ is the sample index, and $x[n]$ is the sampled value of the signal at the nth position.

A.2.1   Summations

Throughout signal processing, we often need to add up long sequences of numbers. While we could easily express this using the standard + notation, such as

$$S = 1 + 2 + 3 + \cdots + 100,$$

this can get cumbersome and difficult to read, especially when the terms of the summation involve other calculations.

Instead, mathematicians express summations using the $\sum$ symbol, derived from the (capital) Greek letter *sigma* (equivalent to S in the latin alphabet, standing for summation). The summation above would be equivalently expressed as

$$S = \sum_{n=1}^{100} n$$
$$= 1 + 2 + 3 + \cdots + 100.$$

The notation specifies the variable which is changing throughout the summation (in our case, n), and the start and end-points of the sum (1 to 100, inclusive) below and above the $\sum$ symbol. Everything after the summation symbol describes what's being added up: in this case, just the values of n.

It's often helpful to think programmatically when reading summation expressions. The summation above would be translated to Python code as follows:

```
S = 0

# Python ranges do not include the end point, so we need to go one␣
↪past
# where we want to stop
for n in range(1, 101):
    S = S + n
```

A.3   MODULAR ARITHMETIC

Another common feature of mathematics of signal processing is the need to deal with repeating sequences and periodicity. We use real numbers to model continuously repeating processes (like a point traveling continuously around a circle), but we also occasionally have **discrete** repeating processes, like the hours on a clock: 12:00, 1:00, 2:00, ..., 11:00. **Modular arithmetic** is the math of discrete repetition.

For a positive integer $N > 1$ (the *modulus*), we define for any integer $q \in \mathbb{Z}$

$$q \mod N \in \{0, 1, \ldots, N-1\}$$

to be the remainder of q/N. Equivalently:

$$k \leftarrow q \mod N \qquad \text{such that } a \cdot N + k = q.$$

for non-negative integer $a \in \mathbb{N}$.

In Python code, this is implemented by

```
# The % symbol
q % N

# Or the numpy function np.mod
np.mod(q, N)
```

A.4 EXAMPLES

A few examples:

- $3 \mod 2 = 1$
- $-1 \mod 4 = 3$
- $20 \mod 5 = 0$
- $10 \mod 10 = 0$

A.5 EXPONENTIALS

An exponential, generally speaking, is an expression of the form c^x for some constant c, and should be read as "c to the xth power". Note that this is different from powers x^n, in that exponentials keep the base c fixed and vary x, while powers do the opposite, as shown in Figure A.1.

Example A.1. $2^x \neq x^2$.

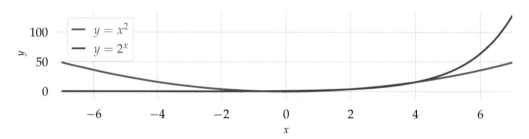

Figure A.1 A quadratic curve $y = x^2$ and an exponential curve $y = 2^x$ behave differently.

Often, when we refer to an *exponential*, we really mean a specific base $e = 2.71828\ldots$, known alternately as "Euler's number", or "the base of the natural logarithm".

A.5.1 Rules for working with exponentials

The following rules (properties) are useful when manipulating expressions involving exponentials. All of these rules work for any base $c \neq 0$, including the special case $c = e$.

Property A.1. Any c raised to the 0 power is 1, and specifically:

$$c^0 = 1$$

Property A.2 (Exponentials turn negatives into inverses:).

$$c^{-a} = \frac{1}{c^a}$$

Property A.3 (Exponentials turn sums into products:).

$$c^a \cdot c^b = c^{a+b}$$

Property A.4 (Exponentials turn products into powers:).

$$c^{a \cdot b} = (c^a)^b = \left(c^b\right)^a$$

A.5.2 The exponential function

While it is convenient to regard e as the number $2.71828\ldots$ there is an alternative (and more general) definition for e given by the following infinite summation

$$e := \sum_{n=0}^{\infty} \frac{1}{n!} \tag{A.1}$$

where

$$n! := 1 \cdot 2 \cdot 3 \cdots n$$

is the *factorial* function: the product of numbers 1 through n. (We define the special case $0! = 1$.)

The exponential function e^x, sometimes also written $\exp(x)$, is equivalent to the number e raised to the xth power as shown in figure A.2.

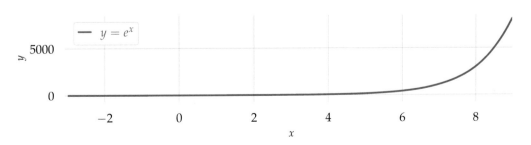

Figure A.2 The exponential function e^x.

Note that summation in (A.1) depends on *infinitely* many terms and is not something you could explicitly compute. While this could be a problem in some cases, it's okay here because the terms diminish quite rapidly:

$$e = \sum_{n=0}^{\infty} \frac{1}{n!} = \frac{1}{0!} + \frac{1}{1!} + \frac{1}{2!} + \frac{1}{3!} + \frac{1}{4!} + \frac{1}{5!} + \dots$$

$$= \frac{1}{1} + \frac{1}{1} + \frac{1}{2} + \frac{1}{6} + \frac{1}{24} + \frac{1}{120} + \dots$$

and the summation *converges* to Euler's number. Proving this sort of thing is out of scope for us here, but we can simulate it computationally by printing out the first few partial sums with a bit of python code.

```python
import numpy as np
from scipy.special import factorial

# Initialize the partial sum
e_partial = 0.0

for n in range(15):
    # Add in 1/n!
    e_partial += 1.0/factorial(n)
    print('First {:2d} term(s): {:0.12f}'.format(n+1, e_partial))

# Print the actual constant to 12 decimals
print('---')
print("Euler's constant: {:0.12f}".format(np.e))
```

```
First  1 term(s): 1.000000000000
First  2 term(s): 2.000000000000
First  3 term(s): 2.500000000000
First  4 term(s): 2.666666666667
First  5 term(s): 2.708333333333
First  6 term(s): 2.716666666667
First  7 term(s): 2.718055555556
First  8 term(s): 2.718253968254
First  9 term(s): 2.718278769841
First 10 term(s): 2.718281525573
First 11 term(s): 2.718281801146
First 12 term(s): 2.718281826198
First 13 term(s): 2.718281828286
First 14 term(s): 2.718281828447
First 15 term(s): 2.718281828458
---
Euler's constant: 2.718281828459
```

After only 15 terms of the summation, we already have a quite good approximation to e (accurate to 11 decimal places).

This definition of e as an infinite summation can be generalized to depend on an arbitrary exponent x:

$$e^x := \sum_{n=0}^{\infty} \frac{x^n}{n!},$$

where the first definition can be recovered by setting $x = 1$ in the second equation.

While we rarely work with this form directly, it is useful because it provides a way to generalize the exponential to support *complex* exponents e^z for $z \in \mathbb{C}$.

A.5.3   Logarithms

Logarithms are the inverse of exponentials, similar to how square-root is the inverse of squaring. In general, logarithms have a "base" $b > 0$, and define the following relationship for input $x > 0$:

$$\log_b x = y \qquad \Leftrightarrow \qquad x = b^y.$$

Intuitively, $\log_b x$ measures the power y we would need to raise b to produce x.

When $x = 0$, there is no finite number y such that $b^y = 0$, and we say $\log_b x = -\infty$.

The base can be any positive number, but among all logarithms, the **natural logarithm** (denoted ln) uses e as the base:

$$\ln x = y \qquad \Leftrightarrow \qquad x = e^y.$$

Other common choices for the base are $b = 2$ and $b = 10$ as shown in Figure A.3. In these cases, $\log_2 x$ and $\log_{10} x$ can be roughly thought of as measuring the "number" of bits (for $b = 2$) or digits (for $b = 10$) needed to represent a number x.

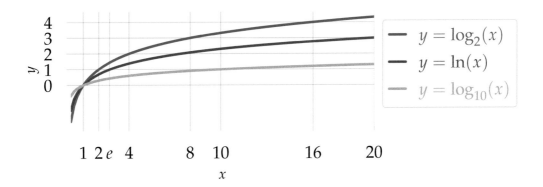

Figure A.3   A comparison of the base-2, natural (base-e), and base-10 logarithms of x.

> Why, exactly, base-e is considered the most "natural" is a deep and fascinating topic, that we unfortunately do not have the space for here.

A.5.4 Properties of logarithms

Property A.5. For any base $b > 0$, $\log_b 1 = 0$.

This is just another way of writing $b^0 = 1$.

Property A.6. For any base $b > 0$, $\log_b b = 1$.

This is another way of saying that $b^1 = b$.

Property A.7. $\log_b x = \frac{\ln x}{\ln b}$

This rule lets us convert between different logarithm bases.

Property A.8. For any x and y, $\log_b(x \cdot y) = \log_b(x) + \log_b(y)$.

This rule allows us to turn products into sums, which can often be computed more easily and with greater precision.

As a corollary, we also get the analogous rule for ratios: $\log_b(x/y) = \log_b(x) - \log_b(y)$.

Property A.9. For any k, $\log_b(x^k) = k \cdot \log_b x$.

This is just a slight generalization of rule 4, allowing us to turn powers into products. Note that k need not be an integer, or positive: it works for any real number!

In particular, we have $\log_b \frac{1}{x} = \log_b \left(x^{-1}\right) = -\log_b x$.

Property A.10. $b^{\log_b x} = x$.

This is just notation that says b^x and $\log_b x$ are mutual inverses.

Bibliography

[AR77]   Jont B Allen and Lawrence R Rabiner. A unified approach to short-time fourier analysis and synthesis. *Proceedings of the IEEE*, 65(11):1558–1564, 1977.

[BT58]   Ralph Beebe Blackman and John Wilder Tukey. The measurement of power spectra from the point of view of communications engineering–part i. *Bell System Technical Journal*, 37(1):185–282, 1958.

[B+30]   Stephen Butterworth and others. On the theory of filter amplifiers. *Wireless Engineer*, 7(6):536–541, 1930.

[CT65]   James W Cooley and John W Tukey. An algorithm for the machine calculation of complex Fourier series. *Mathematics of computation*, 19(90):297–301, 1965.

[EP15]   F. Alton Everest and Ken C. Pohlmann. *Master Handbook of Acoustics*. McGraw-Hill, New York, 6th ed edition, 2015. ISBN 978-0071841047.

[FG66]   James L Flanagan and Roger M Golden. Phase vocoder. *Bell system technical Journal*, 45(9):1493–1509, 1966.

[Fou22] Joseph Fourier. *Theorie analytique de la chaleur, par M. Fourier*. chez Firmin Didot, pere et fils, 1822.

[Kot33] Vladimir Aleksandrovich Kotelnikov. On the transmission capacity of the ether and of cables in electrical communications. In *Proceedings of the first All-Union Conference on the technological reconstruction of the communications sector and the development of low-current engineering*. Moscow, 1933.

[Lyo04] Richard G Lyons. *Understanding Digital Signal Processing*. Prentice Hall PTR, 2004.

[Mul15] Meinard Müller. *Fundamentals of Music Processing*. Springer, Switzerland, 2015. ISBN 9783319219455, 3319219456.

[Nyq28] Harry Nyquist. Certain topics in telegraph transmission theory. *Transactions of the American Institute of Electrical Engineers*, 47(2):617–644, 1928.

[Opp10] Alan V. 1937- Oppenheim. *Discrete-Time Signal Processing*. Pearson, Upper Saddle River, NJ, 2010. ISBN 9780131988422, 0131988425.

[PM72] T Parks and James McClellan. Chebyshev approximation for nonrecursive digital filters with linear phase. *IEEE Transactions on Circuit Theory*, 19(2):189–194, 1972.

[Rem34] Evgeny Y Remez. Sur la détermination des polynômes d'approximation de degré donnée. *Comm. Soc. Math. Kharkov*, 10(196):41–63, 1934.

[Sha49] Claude Elwood Shannon. Communication in the presence of noise. *Proceedings of the IRE*, 37(1):10–21, 1949.

[VHW03] Julius Von Hann and Robert DeCourcy Ward. *Handbook of Climatology*. MacMillan, 1903.

[Whi15] Edmund Taylor Whittaker. On the functions which are represented by the expansions of the interpolation-theory. *Proceedings of the Royal Society of Edinburgh*, 35:181–194, 1915.

Index